Essentials of Astrophysics

Essentials of Astrophysics

Evan Green

NY Research Press
New York

Published by NY Research Press
118-35 Queens Blvd., Suite 400,
Forest Hills, NY 11375, USA
www.nyresearchpress.com

Essentials of Astrophysics
Evan Green

International Standard Book Number: 978-1-63238-891-9 (Hardback)

Cataloging-in-Publication Data

Essentials of astrophysics / Evan Green.
p. cm.
Includes bibliographical references and index.
ISBN 978-1-63238-891-9
1. Astrophysics. 2. Astronomy. 3. Physics. 4. Cosmic physics. I. Green, Evan.
QB461 .E77 2022
523.01--dc23

Contents

Preface

Astrophysics is a branch of astronomy which applies the principles of physics and chemistry. It seeks to determine the nature of astronomical objects, instead of their positions or movement in space. There are several objects studied under this discipline such as stars, galaxies, extrasolar planets and cosmic microwave background. It aims to examine various properties of these objects, such as luminosity, density, temperature and chemical composition, as well as their emissions. There are two major divisions within this field- observational astrophysics and theoretical astrophysics. The former focuses on recording data, and the latter deals with finding out the measurable implications of physical models. The extensive content of this book provides the readers with a thorough understanding of astrophysics. It elucidates the concepts and innovative models around prospective developments with respect to astrophysics. It aims to serve as a resource guide for students and experts alike and contribute to the growth of the discipline.

To facilitate a deeper understanding of the contents of this book a short introduction of every chapter is written below:

Chapter 1- Astrophysics is the branch of science that combines the laws of physics and chemistry to define the nature of astronomical objects. It deals with various principles such as celestial mechanics, Kepler's laws, astrophysical plasma, astrometry, astrophysical magnetohydrodynamics, computational astrophysics, etc. All these diverse principles of astrophysics have been carefully analyzed in this chapter.

Chapter 2- Stellar astrophysics is one of the sub-disciplines of astrophysics which studies the structure and properties of stars. It includes star formation, stellar luminosities, stellar parallax, stellar positions, stellar magnetic fields, nova spectra, etc. This chapter explores these aspects of stellar astrophysics to provide an extensive understanding of the subject.

Chapter 3- The field of science that specializes in the study of the sun is referred to as solar physics. Solar cycle, solar-terrestrial effects, solar magnetic fields, solar wind, solar flares, magnetic flux emergence, etc. fall under its domain. The topics elaborated in this chapter will help in gaining a better perspective about solar physics.

Chapter 4- A black hole is a region in the space which exhibits very strong gravitational acceleration. There are diverse concepts related to black hole such as physics of black holes, black hole entropy, black hole thermodynamics, astrophysics of black holes, creation of a black hole, etc. This chapter has been carefully written to provide an easy understanding of these aspects of black hole.

Chapter 5- Astronomical tools and instruments are used to perform several astronomical observations and calculations. They include optical telescope, solar telescope, radio telescope, hubble space telescope, etc. This chapter closely examines these astronomical tools and instruments to provide an easy understanding of the subject.

I owe the completion of this book to the never-ending support of my family, who supported me throughout the project.

Evan Green

1

Understanding Astrophysics

Astrophysics is the branch of science that combines the laws of physics and chemistry to define the nature of astronomical objects. It deals with various principles such as celestial mechanics, Kepler's laws, astrophysical plasma, astrometry, astrophysical magnetohydrodynamics, computational astrophysics, etc. All these diverse principles of astrophysics have been carefully analyzed in this chapter.

Astrophysics is a branch of space science that applies the laws of physics and chemistry to explain the birth, life and death of stars, planets, galaxies, nebulae and other objects in the universe. It has two sibling sciences, astronomy and cosmology, and the lines between them blur.

In the most rigid sense:

- Astronomy measures positions, luminosities, motions and other characteristics.
- Astrophysics creates physical theories of small to medium-size structures in the universe.
- Cosmology does this for the largest structures, and the universe as a whole.

In practice, the three professions form a tight-knit family. Ask for the position of a nebula or what kind of light it emits, and the astronomer might answer first. Ask what the nebula is made of and how it formed and the astrophysicist will pipe up. Ask how the data fit with the formation of the universe, and the cosmologist would probably jump in. But watch out — for any of these questions, two or three may start talking at once.

While astronomy is one of the oldest sciences, theoretical astrophysics began with Isaac Newton. Prior to Newton, astronomers described the motions of heavenly bodies using complex mathematical models without a physical basis. Newton showed that a single theory simultaneously explains the orbits of moons and planets in space and the trajectory of a cannonball on Earth. This added to the body of evidence for the (then) startling conclusion that the heavens and Earth are subject to the same physical laws.

Perhaps what most completely separated Newton's model from previous ones is that it is predictive as well as descriptive. Based on aberrations in the orbit of Uranus, astronomers predicted the position of a new planet, which was then observed and named Neptune. Being predictive as well as descriptive is the sign of a mature science, and astrophysics is in this category.

Milestones in Astrophysics

Because the only way we interact with distant objects is by observing the radiation they emit, much of astrophysics has to do with deducing theories that explain the mechanisms that produce this radiation, and provide ideas for how to extract the most information from it. The first ideas about the nature of stars emerged in the mid-19th century from the blossoming science of spectral analysis, which means observing the specific frequencies of light that particular substances absorb and emit when heated. Spectral analysis remains essential to the triumvirate of space sciences, both guiding and testing new theories.

Early spectroscopy provided the first evidence that stars contain substances also present on Earth. Spectroscopy revealed that some nebulae are purely gaseous, while some contain stars. This later helped cement the idea that some nebulae were not nebulae at all — they were other galaxies.

In the early 1920s, Cecilia Payne discovered, using spectroscopy, that stars are predominantly hydrogen (at least until their old age). The spectra of stars also allowed astrophysicists to determine the speed at which they move toward or away from Earth. Just like the sound a vehicle emits is different moving toward us or away from us, because of the Doppler shift, the spectra of stars will change in the same way. In the 1930s, by combining the Doppler shift and Einstein's theory of general relativity, Edwin Hubble provided solid evidence that the universe is expanding. This is also predicted by Einstein's theory, and together form the basis of the Big Bang Theory.

Also in the mid-19th century, the physicists Lord Kelvin and Gustav Von Helmholtz speculated that gravitational collapse could power the sun, but eventually realized that energy produced this way would only last 100,000 years. Fifty years later, Einstein's famous $E=mc^2$ equation gave astrophysicists the first clue to what the true source of energy might be (although it turns out that gravitational collapse does play an important role). As nuclear physics, quantum mechanics and particle physics grew in the first half of the 20th century, it became possible to formulate theories for how nuclear fusion could power stars. These theories describe how stars form, live and die, and successfully explain the observed distribution of types of stars, their spectra, luminosities, ages and other features.

Astrophysics is the physics of stars and other distant bodies in the universe, but it also hits close to home. According to the Big Bang Theory, the first stars were almost entirely hydrogen. The nuclear fusion process that energizes them smashes together hydrogen atoms to form the heavier element helium. In 1957, the husband-and-wife astronomer team of Geoffrey and Margaret Burbidge, along with physicists William Alfred Fowler and Fred Hoyle, showed how, as stars age, they produce heavier and heavier elements, which they pass on to later generations of stars in ever-greater quantities. It is only in the final stages of the lives of more recent stars that the elements making up the Earth, such as iron (32.1 percent), oxygen (30.1 percent), silicon (15.1 percent), are produced. Another of these elements is carbon, which together with oxygen, make up the bulk of the mass of all living things, including us. Thus, astrophysics tells us that, while we are not all stars, we are all stardust.

Celestial Mechanics

Celestial mechanics, in the broadest sense, is the application of classical mechanics to the motion of celestial bodies acted on by any of several types of forces. By far the most important force experienced by these bodies, and much of the time the only important force, is that of their mutual gravitational attraction. But other forces can be important as well, such as atmospheric drag on artificial satellites, the pressure of radiation on dust particles, and even electromagnetic forces on dust particles if they are electrically charged and moving in a magnetic field.

The term celestial mechanics is sometimes assumed to refer only to the analysis developed for the motion of point mass particles moving under their mutual gravitational attractions, with emphasis on the general orbital motions of solar system bodies. The term astrodynamics is often used to refer to the celestial mechanics of artificial satellite motion. Dynamic astronomy is a much broader term, which, in addition to celestial mechanics and astrodynamics, is usually interpreted to include all aspects of celestial body motion (e.g., rotation, tidal evolution, mass and mass distribution determinations for stars and galaxies, fluid motions in nebulas, and so forth).

Perturbations and Problems of Two Bodies

The Approximate Nature of Kepler's Laws

The constraints placed on the force for Kepler's laws to be derivable from Newton's laws were that the force must be directed toward a central fixed point and that the force must decrease as the inverse square of the distance. In actuality, however, the

Sun, which serves as the source of the major force, is not fixed but experiences small accelerations because of the planets, in accordance with Newton's second and third laws. Furthermore, the planets attract one another, so that the total force on a planet is not just that due to the Sun; other planets perturb the elliptical motion that would have occurred for a particular planet if that planet had been the only one orbiting an isolated Sun. Kepler's laws therefore are only approximate. The motion of the Sun itself means that, even when the attractions by other planets are neglected, Kepler's third law must be replaced by $(M + mi)\tau^2 \propto a^3$, where mi is one of the planetary masses and M is the Sun's mass. That Kepler's laws are such good approximations to the actual planetary motions results from the fact that all the planetary masses are very small compared to that of the Sun. The perturbations of the elliptic motion are therefore small, and the coefficient $M + m_i \approx M$ for all the planetary masses mi means that Kepler's third law is very close to being true.

Newton's second law for a particular mass is a second-order differential equation that must be solved for whatever forces may act on the body if its position as a function of time is to be deduced. The exact solution of this equation, which resulted in a derived trajectory that was an ellipse, parabola, or hyperbola, depended on the assumption that there were only two point particles interacting by the inverse square force. Hence, this "gravitational two-body problem" has an exact solution that reproduces Kepler's laws. If one or more additional bodies also interact with the original pair through their mutual gravitational interactions, no exact solution for the differential equations of motion of any of the bodies involved can be obtained. However, the motion of a planet is almost elliptical, since all masses involved are small compared to the Sun. It is then convenient to treat the motion of a particular planet as slightly perturbed elliptical motion and to determine the changes in the parameters of the ellipse that result from the small forces as time progresses. It is the elaborate developments of various perturbation theories and their applications to approximate the exact motions of celestial bodies that has occupied celestial mechanicians since Newton's time.

Perturbations of Elliptical Motion

So far the following orbital parameters, or elements, have been used to describe elliptical motion: the orbital semimajor axis a, the orbital eccentricity e, and, to specify position in the orbit relative to the perihelion, either the true anomaly f, the eccentric anomaly u, or the mean anomaly l. Three more orbital elements are necessary to orient the ellipse in space, since that orientation will change because of the perturbations. The most commonly chosen of these additional parameters are illustrated in figure, where the reference plane is chosen arbitrarily to be the plane of the ecliptic, which is the plane of Earth's orbit defined by the path of the Sun on the sky. (For motion of a near-Earth artificial satellite, the most convenient reference plane would be that of Earth's Equator.) Angle i is the inclination of the orbital plane to the reference plane. The line of nodes is the intersection of the orbit plane with the reference plane, and the ascending node is that point where the planet travels from below the reference plane (south) to above the reference plane (north). The ascending node is described by its angular position measured from a reference point on the ecliptic plane, such as the vernal equinox; the angle Ω is called the longitude of the ascending node. Angle ω (called the argument of perihelion) is the angular distance from the ascending node to the perihelion measured in the orbit plane.

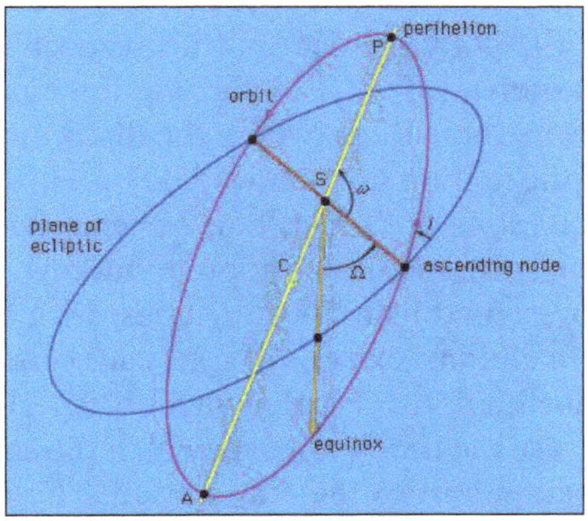

Orbital elements Ω, ω, and i orienting the ellipse.

For the two-body problem, all the orbital parameters a, e, i, Ω, and ω are constants. A sixth constant T, the time of perihelion passage (i.e., any date at which the object in orbit was known to be at perihelion), may be used to replace f, u, or l, and the position of the planet in its fixed elliptic orbit can be determined uniquely at subsequent times. These six constants are determined uniquely by the six initial conditions of three components of the position vector and three components of the velocity vector relative to a coordinate system that is fixed with respect to the reference plane. When small perturbations are taken into account, it is convenient to consider the orbit as an instantaneous ellipse whose parameters are defined by the instantaneous values of the position and velocity vectors, since for small perturbations the orbit is approximately an ellipse. In fact, however, perturbations cause the six formerly constant parameters to vary slowly, and the instantaneous perturbed orbit is called an osculating ellipse; that is, the osculating ellipse is that elliptical orbit that would be assumed by the body if all the perturbing forces were suddenly turned off.

First-order differential equations describing the variation of the six orbital parameters can be constructed for each planet or other celestial body from the second-order differential equations that result by equating the mass times the acceleration of a body to the sum of all the forces acting on the body (Newton's second law). These equations are sometimes called the Lagrange planetary equations after their derivation by the great Italian-French mathematician Joseph-Louis Lagrange. As long as the forces are conservative and do not depend on the velocities—i.e., there is no loss of mechanical energy through such processes as friction—they can be derived from partial derivatives of a function of the spatial coordinates only, called the potential energy, whose magnitude depends on the relative separations of the masses.

In the case where all the forces are derivable from such potential energy, the total energy of a system of any number of particles—i.e., the kinetic energy plus the potential energy—is constant. The kinetic energy of a single particle is one-half its mass times the square of its velocity and the total kinetic energy is the sum of such expressions for all the particles being considered. The conservation of energy principle is thus expressed by an equation relating the velocities of all the masses to their positions at any time. The partial derivatives of the potential energy with respect to spatial coordinates are transformed into particle derivatives of a disturbing function with respect to the orbital elements in the Lagrange equations, where the disturbing function vanishes if all bodies perturbing the elliptic motion are removed. Like Newton's equations of motion, Lagrange's differential equations are exact, but they can be solved only numerically on a computer or analytically by successive approximations. In the latter process, the disturbing function is represented by a Fourier series, with convergence of the series (successive decrease in size and importance of the terms) depending on the size of the orbital eccentricities and inclinations. Clever changes of variables and other mathematical tricks are used to increase the time span over which the solutions (also represented by series) are good approximations to the real motion. These series solutions usually diverge, but they still represent the actual motions remarkably well for limited periods of time. One of the major triumphs of celestial mechanics using these perturbation techniques was the discovery of Neptune in 1846 from its perturbations of the motion of Uranus.

Examples of Perturbations

Some of the variations in the orbital parameters caused by perturbations can be understood in simple terms. The lunar orbit is inclined to the ecliptic plane by about 5°, and the longitude of its ascending node on the ecliptic plane (Ω in figure above) is observed to regress (Ω decreasing) a complete revolution in 18.61 years. The Sun is the dominant cause of this regression of the lunar node. When the Moon is closer to the Sun than Earth, the Sun accelerates the Moon slightly more than it accelerates Earth. This difference in the accelerations is what perturbs the lunar motion around Earth. The Moon does not fly off in this situation, since the acceleration of the Moon toward Earth is much larger than the difference between the Sun's accelerations of Earth and the Moon.

The Sun, of course, is always in the ecliptic plane, since its apparent path among the stars defines the plane. This means that the perturbing acceleration just defined will always be pointed slightly toward the ecliptic plane whenever the Moon is below or above this plane in its orbital motion about Earth. This tendency to pull the Moon toward the ecliptic plane means that the Moon will cross the plane on each half orbit at a longitude that is slightly smaller than the longitude at which it would have crossed if the Sun had not been there. Thus, the line of nodes will have regressed. The instantaneous rate at which the node regresses varies as the geometry changes during the Moon's motion around Earth, and during the Earth-Moon system's motion around the Sun, but there is always a net regression. Such a change that is always in the same direction as time increases is called a secular perturbation. Superposed on the secular perturbation of the longitude of the node are periodic perturbations (periodically changing their direction), which are revealed by the fact that the rate of secular regression of the node is not constant in time. The Sun causes a secular increase in the longitude of the lunar perigee ($\Omega + \omega$ in figure above) of one complete revolution in 8.85 years, as well as periodic perturbations in the inclination, eccentricity, and mean motion.

For near-Earth artificial satellites, the deviation of Earth's mass distribution from spherical symmetry is the dominant cause of the perturbations from pure elliptic motion. The most important deviation is the equatorial bulge of Earth due to its rotation. If, for example, Earth were a sphere with a ring of mass around its Equator, the ring would give to a satellite whose orbit is inclined to the Equator a component of acceleration toward the Equator plane whenever the satellite was above or below this plane. By an argument similar to that for the Moon acted on by the Sun, this acceleration would cause the line of nodes of a close satellite orbit to regress a little more than 5° per day.

As a final example, the distribution of continents and oceans and the varying mass densities in Earth's mantle (the layer underlying the crust) lead to a slight deviation of Earth's gravitational force field from axial symmetry. Usually these causes only short-period perturbations of low amplitude for near-Earth satellites. However, communications or weather satellites that are meant to maintain a fixed longitude over the Equator (i.e., geostationary satellites, which orbit synchronously with Earth's rotation) are destabilized by this deviation except at two longitudes. If the axial asymmetry is represented by a slightly elliptical Equator, the difference between the major and minor axis of the ellipse is about 64 metres, with the major axis located about 35° W. A satellite at a position slightly ahead of the long axis of the elliptical Equator will experience a component of acceleration opposite its direction of orbital motion (as if a large mountain were pulling it back). This acceleration makes the satellite fall closer to Earth and increases its mean motion, causing it to drift further ahead of the axial bulge on the Equator. If the satellite is slightly behind the axial bulge, it experiences acceleration in the direction of its motion. This makes the satellite move away from Earth with a decrease in its mean motion, so that it will drift further behind the axial bulge. The synchronous Earth satellites are thus repelled from the long axis of the equatorial ellipse and attracted to the short axis, and compensating accelerations, usually from onboard jets, are required to stabilize a satellite at any longitude other than the two corresponding to the ends of the short axis of the axial bulge. (The jets are actually required for any longitude, as they must also compensate for other perturbations such as radiation pressure).

The Three-body Problem

The inclusion of solar perturbations of the motion of the Moon results in a "three-body problem" (Earth-Moon-Sun), which is the simplest complication of the completely solvable two-body problem. When Earth, the Moon, and the Sun are considered to be point masses, this particular three-body problem is called "the main problem of the lunar theory," which has been studied extensively with a variety of methods beginning with Newton. Although the three-body problem has no complete analytic solution in closed form, various series solutions by successive approximations achieve such accuracy that complete theories of the lunar motion must include the effects of the nonspherical mass distributions of both Earth and the Moon as well as the effects of the planets if the precision of the predicted positions is to approach that of the observations. Most of the schemes for the main problem are partially numerical and therefore apply only to the lunar motion. An exception is the completely analytic work of the French astronomer Charles-Eugène Delaunay, who exploited and developed the most elegant techniques of classical mechanics pioneered by his contemporary, the Irish astronomer and mathematician William R. Hamilton. Delaunay could predict the position of the Moon to within its own diameter over a 20-year time span. Since his development was entirely analytic, the work was applicable to the motions of satellites about other planets where the series expansions converged much more rapidly than they did for the application to the lunar motion.

Delaunay's work on the lunar theory demonstrates some of the influence that celestial mechanics has had on the development of the techniques of classical mechanics. This close link between the development of classical mechanics and its application to celestial mechanics was probably no better demonstrated than in the work of the French mathematician Henri Poincaré. Poincaré, along with other great mathematicians such as George D. Birkhoff, Aurel Wintner, and Andrey N. Kolmogorov, placed celestial mechanics on a sounder mathematical basis and was less concerned with quantitatively accurate prediction of celestial body motion. Poincaré demonstrated that the series solutions in use in celestial mechanics for so long generally did not converge but that they could give accurate descriptions of the motion for significant periods of time in truncated form. The elaborate theoretical developments in celestial and classical mechanics have received more attention recently with the realization that a large class of motions are of an irregular or chaotic nature and require fundamentally different approaches for their description.

The Restricted Three-body Problem

The simplest form of the three-body problem is called the restricted three-body problem, in which a particle of infinitesimal mass moves in the gravitational field of two massive bodies orbiting according to the exact solution of the two-body problem. (The particle with infinitesimal mass, sometimes called a massless particle, does not perturb the motions of the two massive bodies). There is an enormous literature devoted to this problem, including both analytic and numerical developments. The analytic work was devoted mostly to the circular, planar restricted three-body problem, where all particles are confined to a plane and the two finite masses are in circular orbits around their centre of mass (a point on the line between the two masses that is closer to the more massive). Numerical developments allowed consideration of the more general problem.

In the circular problem, the two finite masses are fixed in a coordinate system rotating at the orbital angular velocity, with the origin (axis of rotation) at the centre of mass of the two bodies. Lagrange showed that in this rotating frame there were five stationary points at which the massless particle would remain fixed if placed there. There are three such points lying on the line connecting the two finite masses: one between the masses and one outside each of the masses. The other two stationary points, called the triangular points, are located equidistant from the two finite masses at a distance equal to the finite mass separation. The two masses and the triangular stationary points are thus located at the vertices of equilateral triangles in the plane of the circular orbit.

There is a constant of the motion in the rotating frame that leads to an equation relating the velocity of the massless particle in this frame to its position. For given values of this constant it is possible to construct curves in the plane on which the velocity vanishes. If such a zero-velocity curve is closed, the particle cannot escape from the interior of the closed zero-velocity curve if placed there with the constant of the motion equal to the value used to construct the curve. These zero-velocity curves can be used to show that the three collinear stationary points are all unstable in the sense that, if the particle is placed at one of these points, the slightest perturbation will cause it to move far away. The triangular points are stable if the ratio of the finite masses is less than 0.04, and the particle would execute small oscillations around one of the triangular points if it were pushed slightly away. Since the mass ratio of Jupiter to the Sun is about 0.001, the stability criterion is satisfied, and Lagrange predicted the presence of the Trojan asteroids at the triangular points of the Sun-Jupiter system 134 years before they were observed. Of course, the stability of the triangular points must also depend on the perturbations by any other bodies. Such perturbations are sufficiently small not to destabilize the Trojan asteroids. Single Trojan-like bodies have also been found orbiting at leading and trailing triangular points in the orbits of Neptune and of Saturn's satellite Tethys, at the leading triangular point in the orbit of another Saturnian satellite, Dione, and at the trailing point in the orbit of Mars.

Orbital Resonances

There are stable configurations in the restricted three-body problem that are not stationary in the rotating frame. If, for example, Jupiter and the Sun are the two massive bodies, these stable configurations occur when the mean motions of Jupiter and the small particle—here an asteroid—are near a ratio of small integers. The orbital mean motions are then said to be nearly commensurate, and an asteroid that is trapped near such mean motion commensurability is said to be in an orbital resonance with Jupiter. For example, the Trojan asteroids librate (oscillate) around the 1:1 orbital resonance (i.e., the orbital period of Jupiter is in a 1:1 ratio with the orbital period of the Trojan asteroids); the asteroid Thule librates around the 4:3 orbital resonance; and several asteroids in the Hilda group librate around the 3:2 orbital resonance. There are several such stable orbital resonances among the satellites of the major planets and one involving Pluto and the planet Neptune. The analysis based on the restricted three-body problem cannot be used for the satellite resonances, however, except for the 4:3 resonance between Saturn's satellites Titan and Hyperion, since the participants in the satellite resonances usually have comparable masses.

Although the asteroid Griqua librates around the 2:1 resonance with Jupiter, and Alinda librates around the 3:1 resonance, the orbital commensurabilities 2:1, 7:3, 5:2, and 3:1 are characterized by an absence of asteroids in an otherwise rather highly populated, uniform distribution spanning all of the commensurabilities. These are the Kirkwood gaps in the distribution of asteroids, and the recent understanding of their creation and maintenance has introduced into celestial mechanics an entirely new concept of irregular, or chaotic, orbits in a system whose equations of motion are entirely deterministic.

Chaotic Orbits

The French astronomer Michel Hénon and the American astronomer Carl Heiles discovered that when a system exhibiting periodic motion, such as a pendulum, is perturbed by an external force that is also periodic, some initial conditions lead to motions where the state of the system becomes essentially unpredictable (within some range of system states) at some time in the future, whereas initial conditions within some other set produce quasiperiodic or predictable behaviour. The unpredictable behaviour is called chaotic, and initial conditions that produce it are said to lie in a chaotic zone. If the chaotic zone is bounded, in the sense that only limited ranges of initial values of the variables describing the motion lead to chaotic behaviour, the uncertainty in the state of the system in the future is limited by the extent of the chaotic zone; that is, values of the variables in the distant future are completely uncertain only within those ranges of values within the chaotic zone. This complete uncertainty within the zone means the system will eventually come arbitrarily close to any set of values of the variables within the zone if given sufficient time. Chaotic orbits were first realized in the asteroid belt.

A periodic term in the expansion of the disturbing function for a typical asteroid orbit becomes more important in influencing the motion of the asteroid if the frequency with which it changes sign is very small and its coefficient is relatively large. For asteroids orbiting near mean motion commensurability with Jupiter, there are generally several terms in the disturbing function with large coefficients and small frequencies that are close but not identical. These "resonant" terms often dominate the perturbations of the asteroid motion so much that all the higher-frequency terms can be neglected in determining a first approximation to the perturbed motion. This neglect is equivalent to averaging the higher-frequency terms to zero; the low-frequency terms change only slightly during the averaging. If one of the frequencies vanishes on the average, the periodic term becomes nearly constant, or secular, and the asteroid is locked into an exact orbital resonance near the particular mean motion commensurability. The mean motions are not exactly commensurate in such a resonance, however, since the motion of the asteroid orbital node or perihelion is always involved (except for the 1:1 Trojan resonances).

For example, for the 3:1 commensurability, the angle $\theta = \lambda_A - 3\lambda_J + \varpi_A$ is the argument of one of the important periodic terms whose variation can vanish (zero frequency). Here $\lambda = \Omega + \omega + l$ is the mean longitude, the subscripts A and J refer to the asteroid and Jupiter, respectively, and $\varpi = \Omega + \omega$ is the longitude of perihelion. Within resonance, the angle θ librates, or oscillates, around a constant value as would a pendulum around its equilibrium position at the bottom of its swing. The larger the amplitude of the equivalent pendulum, the larger its velocity at the bottom of its swing. If the velocity of the pendulum at the bottom of its swing, or, equivalently, the maximum rate of change of the angle θ, is sufficiently high, the pendulum will swing over the top of its support and be in a state of rotation instead of libration. The maximum value of the rate of change of θ for which θ remains an angle of libration (periodically reversing its variation) instead of one of rotation (increasing or decreasing monotonically) is defined as the half-width of the resonance.

Another term with nearly zero frequency when the asteroid is near the 3:1 commensurability has the argument $\theta' = \lambda_A - \lambda_J + 2\varpi_J$. The substitution of the longitude of Jupiter's perihelion for that of the asteroid means that the rates of change of θ and θ' will be slightly different. As the resonances are not separated much in frequency, there may exists values of the mean motion of the asteroid where both θ and θ' would be angles of libration if either resonance existed in the absence of the other. The resonances are said to overlap in this case, and the attempt by the system to librate simultaneously about both resonances for some initial conditions leads to chaotic orbital behaviour. The important characteristic of the chaotic zone for asteroid motion near mean motion commensurability with Jupiter is that it includes a region where the asteroid's orbital eccentricity is large. During the variation of the elements over the entire chaotic zone as time increases, large eccentricities must occasionally be reached. For asteroids near the 3:1 commensurability with Jupiter, the orbit then crosses that of Mars, whose gravitational interaction in a close encounter can remove the asteroid from the 3:1 zone.

By numerically integrating many orbits whose initial conditions spanned the 3:1 Kirkwood gap region in the asteroid belt, Jack Wisdom, an American dynamicist who developed a powerful means of analyzing chaotic motions, found that the chaotic zone around this gap precisely matched the physical extent of the gap. There are no observable asteroids with orbits within the chaotic zone, but there are many just outside extremes of the zone. Other Kirkwood gaps can be similarly accounted for. The realization that orbits governed by Newton's laws of motion and gravitation could have chaotic properties and that such properties could solve a long-standing problem in the celestial mechanics of the solar system is a major breakthrough in the subject.

The N-body Problem

The general problem of n bodies, where n is greater than three, has been attacked vigorously with numerical techniques on powerful computers. Celestial mechanics in the solar system is ultimately an n-body problem, but the special configurations and relative smallness of the perturbations have allowed quite accurate descriptions of motions (valid for limited time periods) with various approximations and procedures without any attempt to solve the complete problem of n bodies. Examples are the restricted three-body problem to determine the effect of Jupiter's perturbations of the asteroids and the use of successive approximations of series solutions to sequentially add the effects of smaller and smaller perturbations for the motion of the Moon. In the general n-body problem, all bodies have arbitrary masses, initial velocities, and positions; the bodies interact through Newton's law of gravitation, and one attempts to determine the subsequent motion of all the bodies. Many numerical solutions for the motion of quite large numbers of gravitating particles have been successfully completed where the precise motion of individual particles is usually less important than the statistical behaviour of the group.

Numerical Solutions

Numerical solutions of the exact equations of motion for n bodies can be formulated. Each body is subject to the gravitational attraction of all the others, and it may be subject to other forces as well. It is relatively easy to write the expression for the instantaneous acceleration (equation of motion) of each body if the position of all the other bodies is known, and expressions for all the other forces can be written (as they can for gravitational forces) in terms of the relative positions of the particles and other defining characteristics of the particle and its environment. Each particle is allowed to move under its instantaneous acceleration for a short time step. Its velocity and position are thereby changed, and the new values of the variables are used to calculate the acceleration for the next time step, and so forth. Of course, the real position and velocity of the particle after each time step will differ from the calculated values by errors of two types. One type results from the fact that the acceleration is not really constant over the time step, and the other from the rounding off or truncation of the numbers at every step of the calculation. The first type of error is decreased by taking shorter time steps. But this means more numerical operations must be carried out over a given span of time, and this increases the round-off error for a given precision of the numbers being carried in the calculation. The design of numerical algorithms, as well as the choice of precisions and step sizes that maximize the speed of the calculation while keeping the errors within reasonable bounds, is almost an art form developed by extensive experience and ingenuity. For example, a scheme exists for extrapolating the step size to zero in order to find the change in the variables over a relatively short time span, thereby minimizing the accumulation of error from this source. If the total energy of the system is theoretically conserved, its evaluation for values of the variables at the beginning and end of a calculation is a measure of the errors that have accumulated.

The motion of the planets of the solar system over time scales approaching its 4.6-billion-year age is a classic n-body problem, where n = 9 with the Sun included. The question of whether or not the solar system is ultimately stable—whether the current configuration of the planets will be maintained indefinitely under their mutual perturbations, or whether one or another planet will eventually be lost from the system or otherwise have its orbit drastically altered—is a long-standing one that might someday be answered through numerical calculation. The interplay of orbital resonances and chaotic orbits can be investigated numerically, and this interplay may be crucial in determining the stability of the solar system. Already it appears that the parameters defining the orbits of several planets vary over narrow chaotic zones, but whether or not this chaos can lead to instability if given enough time is still uncertain.

If accelerations are determined by summing all the pairwise interactions for the n particles, the computer time per time step increases as n^2. Practical computations for the direct calculation of the interactions between all the particles are thereby limited to $n < 10{,}000$. Therefore, for larger values of n, schemes are used where a particle is assumed to move in the force field of the remaining particles approximated by that due to a continuum mass distribution, or a "tree structure" is used where the effects of nearby particles are considered individually while larger and larger groups of particles are considered collectively as their distances increase. These later schemes have the capability of calculating the evolution of a very large system of particles using a reasonable amount of computer time with reasonable approximation. Values of n near 100,000 have been used in calculations determining the evolution of galaxies of stars. Values of n in the billions have been used in calculations of galaxy formation in the early universe. The consequences for distribution of stars when two galaxies closely approach one another or even collide has been determined. Even calculations of the n-body problem where n changes with time have been completed in the study of the accumulation of larger bodies from smaller bodies via collisions in the process of the formation of the planets.

In all n-body calculations, very close approaches of two particles can result in accelerations so large and so rapidly changing that large errors are introduced or the calculation completely diverges. Accuracy can sometimes be maintained in such a close approach, but only at the expense of requiring very short time steps, which drastically slows the calculation. When n is small, as in some solar system calculations where two-body orbits still dominate, close approaches are sometimes handled by a change to a set of variables, usually involving the eccentric anomaly u, that vary much less rapidly during the encounter. In this process, called regularization, the encounter is traversed in less computer time while preserving reasonable accuracy. This process is impractical when n is large, so accelerations are usually artificially bounded on close approaches to prevent instabilities in the numerical calculation and to prevent slowing the calculation. For example, if several sets of particles were trapped in stable, close binary orbits, the very short time steps required to follow this rapid motion would bring the calculation to a virtual standstill, and such binary motion is not important in the overall evolution of, say, a galaxy of stars.

Algebraic Maps

In numerical calculations for conservative systems with modest values of n over long time spans, such as those seeking a determination of the stability of the solar system, the direct solution of the differential equations governing the motions requires excessive time on any computer and accumulates excessive round-off error in the process. Excessive time also is required to explore thoroughly a complete range of orbital parameters in numerical experiments in order to determine the extent of chaotic zones in various configurations (e.g., those in the asteroid belt near orbital mean motion commensurabilities with Jupiter). A solution to this problem is the use of an algebraic map, which maps the space of system variables onto itself in such a way that the values of all the variables at one instant of time are expressed by algebraic relations in terms of the values of the variables at a fixed time in the past. The values at the next time step are determined by applying the same map to the values just obtained, and so on. The map is constructed by assuming that the motions of all the bodies are unperturbed for a given short time but are periodically "kicked" by the perturbing forces for only an instant. The continuous perturbations are thus replaced by periodic impulses. The values of the variables are "mapped" from one time step to the next by the fact that the unperturbed part of the motion is available from the exact solution of the two-body problem, and it is easy to solve the equations with all the perturbations over the short time of the impulse. Although this approximation does not produce exactly the same values of all the variables at some time in the future as those produced by a numerical solution of the differential equations starting with the same initial conditions, the qualitative behaviour is indistinguishable over long time periods. As computers can perform the algebraic calculations as much as 1,000 times faster than they can solve the corresponding differential equations, the computational time savings are enormous and problems otherwise impossible to explore become tractable.

Tidal Evolution

This discussion has so far treated the celestial mechanics of bodies accelerated by conservative forces (total energy being conserved), including perturbations of elliptic motion by nonspherical mass distributions of finite-size bodies. However, the gravitational field of one body in close orbit about another will tidally distort the shape of the other body. Dissipation of part of the energy stored in these tidal distortions leads to a coupling that causes secular changes (always in the same direction) in the orbit and in the spins of both bodies. Since tidal dissipation accounts for the current spin states of several planets, the spin states of most of the planetary satellites and some of their orbital configurations, and the spins and orbits of close binary stars.

The twice-daily high and low tides in the ocean are known by all who have lived near a coast. Few are aware, however, that the solid body of Earth also experiences twice-daily tides with a maximum amplitude of about 30 centimetres. George Howard Darwin, the second son of Charles Darwin, the naturalist, was an astronomer-geophysicist who understood quantitatively the generation of the tides in the gravitational fields of tide-raising bodies, which are primarily the Moon and the Sun for Earth; he pointed out that the dissipation of tidal energy resulted in a slowing of Earth's rotation while the Moon's orbit was gradually expanded. That any mass raises a tide on every other mass within its gravitational field follows from the fact that the gravitational force between two masses decreases as the inverse square of the distance between them.

The accelerations due to mass m_s of three mass elements in the spherical mass m_p are proportional to the length of the arrows attached to each element. The element nearest m_s is accelerated more than the element at the centre of m_p and tends to leave the centre element behind; the element at the centre of m_p is accelerated more than the element farthest from m_s, and the latter tends to be left further behind. The point of view of a fictitious observer at the centre of m_p can be realized by subtracting the acceleration of the central mass element from that of each of the other two mass elements. If the mass elements were free, this observer would see the two extreme mass elements being accelerated in opposite directions away from his position at the centre.

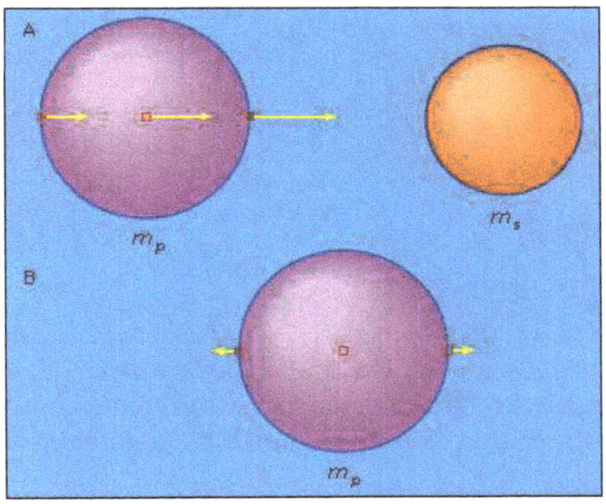

Variation of gravitational acceleration across a finite-sized body leading to differential acceleration relative to its centre.

But the mass elements are not free; they are gravitationally attracted to one another and to the remaining mass in m_p. The gravitational acceleration of the mass elements on the surface of m_p toward the centre of m_p far exceeds the differential acceleration due to the gravitational attraction of m_s, thus the elements do not fly off. If m_p were incompressible and perfectly rigid, the mass elements on the surface would weigh a little less than they would if m_s were not there but would not move relative to the centre of m_p. If m_p were fluid or otherwise not rigid, it would distort into an oval shape in the presence of m_s. The reason for this distortion is that the mass elements making up m_p that do not lie on the line joining the centres of m_p and m_s also experience a differential acceleration. Such differential accelerations are not perpendicular to the surface, however, and are therefore not compensated by the self-gravity that accelerates mass elements toward the centre of m_p. This is shown in figure, where one of the differential accelerations is resolved into two components (dotted arrows), one perpendicular and one tangential to the surface. The perpendicular component is compensated by the self-gravity; the tangential component is not. If m_p were entirely fluid, the uncompensated tangential components of the differential accelerations due to m_s would cause mass to flow toward the points on m_p that were either closest to m_s or farthest from m_s until m_p would resemble figure. In this shape the self-gravity is no longer perpendicular to the surface but has a component opposite the tangential component of the differential acceleration. Only in this distorted shape will all the differential accelerations be compensated and the entire body accelerated like the centre. If m_p is not fluid but is rigid like rock or iron, part of the compensating acceleration will be provided by internal stress forces, and the body will distort less. As no material is perfectly rigid, there is always some tidal bulge, and compressibility of the material will farther enhance this bulge. Note that the tidal distortion is independent of the orbital motion and would also occur if m_p and m_s were simply falling toward each other. (There is a similar tide raised on m_s by m_p that will be ignored for the present).

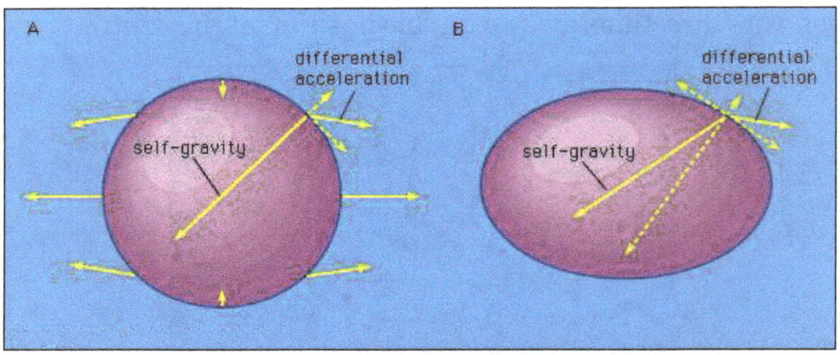

Figure: (A) Uncompensated tangential accelerations cause (B) tidal distortions in which all the differential accelerations are balanced by the change in the self-gravitational acceleration resulting from the distortion and by internal stresses.

If m_p rotates relative to m_s, an observer on the surface of m_p would successively rotate through the maxima and minima of the tidal distortion, which would tend to remain aligned with the direction to m_s. The observer would thereby experience two high and two low tides a day, as observed on Earth. Some of the energy of motion of any fluid parts of m_p and some of the energy stored as distortion of the solid parts as the tides wax and wane is converted into heat, and this dissipation of mechanical energy causes a delay in the response of the body to the tide-raising force. This means that high tide would occur at a given point on m_p as it rotates relative to m_s after m_s passes overhead. (On Earth, the continents alter the motion of the fluid ocean so much that ocean tides at continental coasts do not always lag the passage of the Moon overhead.) If m_p rotates in the same direction as m_s revolves in its orbit, the tidal bulge is carried ahead of m_s, as shown in figure by angle δ. Again, because the gravitational force between two masses varies as the inverse square of their separation, the tidal bulge closest to m_s experiences a greater attraction toward m_s than does the bulge farthest away (F_2). As these two forces are not aligned with the centre of m_p, there is a twisting effect, or torque, on mp that retards its rate of rotation. This retardation will continue until the rotation is synchronous with the mean orbital motion of m_s. This has happened for the Moon, which keeps the same face toward Earth.

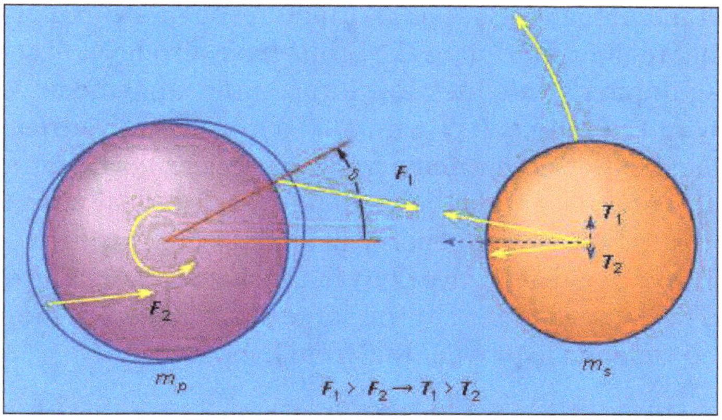

Unequal forces on two tidal bulges, leading to retardation of the spin of m and an acceleration of m in its orbit.

From Newton's third law, there are equal and opposite forces acting on m_s corresponding to F_1 and F_2. In figures these forces are represented as T_1 and T_2, and each has been resolved into two components, one directed toward the centre of m_p and the other perpendicular to this direction. The inequality of these forces causes a net acceleration of m_s in its orbit, which thereby expands, as is observed for the Moon. Both the observed increase in the length of one day of 0.0016 second per century and the observed recession of the Moon of 3 to 4 centimetres per year are understood as consequences of the tides raised on Earth.

In figure, it has been assumed that the spin axis of m_p is perpendicular to the plane of the orbit of m_s. If the spin axis is inclined to this plane, the tidal bulge is carried out of the plane as well as ahead of m_s. This means that there is a twist, or torque, that changes the direction of the spin axis, so both the magnitude of spin and the direction of the spin axis experience a tidal evolution. The end point of tidal evolution for the spin state of one body of an isolated pair is rotation synchronous with the mean orbital motion with the spin axis perpendicular to the orbit plane. This simple picture is complicated somewhat if other perturbations cause the orbital plane to process. This precession for the lunar orbit causes its spin axis to be inclined 6°41′ to the orbit plane as the end point of tidal evolution.

In addition to those of the Earth-Moon pair, numerous other consequences of tidal dissipation and the resulting evolution can be observed in the solar system and elsewhere in the Milky Way Galaxy. For example, all the major and close planetary satellites but one is observed to be rotating synchronously with their orbital motion. The exception is Saturn's satellite Hyperion. Tidal friction has indeed retarded Hyperion's initial spin rate to a value near that of synchronous rotation, but the combination of Hyperion's unusually asymmetric shape and its high orbital eccentricity leads to gravitational torques that make synchronous rotation unstable. As a result, the tides have brought Hyperion to a state where it tumbles chaotically with large changes in the direction and magnitude of its spin on time scales comparable to its orbital period of about 21 days.

The assembly and maintenance of several orbital resonances among the satellites because of differential tidal expansion of the orbits have also been observed. The orbital resonances among Jupiter's satellites Io, Europa, and Ganymede, where the orbital periods are nearly in the ratio 1:2:4, maintain Io's orbital eccentricity at the value of 0.0041. This rather modest eccentricity causes sufficient variation in the magnitude and direction of Io's enormous tidal bulge to have melted a significant fraction of the satellite through dissipation of tidal energy in spite of Io's synchronous rotation. As a result, Io is the most volcanically active body in the solar system. The orbital eccentricity would normally be damped to zero by this large dissipation, but the orbital resonances with Europa and Ganymede prevent this from happening.

The distant dwarf planet Pluto and its satellite Charon have almost certainly reached the ultimate end point where further tidal evolution has ceased altogether (the tiny tides raised by the Sun and other planets being neglected). In this state the orbit is circular, with both bodies rotating synchronously with the orbital motion and both spin axes perpendicular to the orbital plane.

The spin of the planet Mercury has been slowed by tides raised by the Sun to a final state where the spin angular velocity is exactly 1.5 times the orbital mean motion. This state is stable against further change because Mercury's high orbital eccentricity (0.206) allows restoring torques on the permanent (nontidal) axial asymmetry of the planet, which keeps the longest equatorial axis aligned with the direction to the Sun at perihelion. The tidal reduction of Mercury's average eccentricity (near 0.2) will cause insufficient change during the remaining lifetime of the Sun to disrupt this spin-orbit resonance. Finally, many close binary stars are observed to have circular orbits and synchronized spins—an example of tidal evolution elsewhere in the Milky Way Galaxy.

Kepler's Laws

Kepler's laws of planetary motion in astronomy and classical physicsa are the laws describing the motions of the planets in the solar system. They were derived by the German astronomer Johannes Kepler, whose analysis of the observations of the 16th-century Danish astronomer Tycho Brahe enabled him to announce his first two laws in the year 1609 and a third law nearly a decade later, in 1618. Kepler himself never numbered these laws or specially distinguished them from his other discoveries.

Kepler's three laws of planetary motion can be stated as follows: (1) All planets move about the Sun in elliptical orbits, having the Sun as one of the foci. (2) A radius vector joining any planet to the Sun sweeps out equal areas in equal lengths of time. (3) The squares of the sidereal periods (of revolution) of the planets are directly proportional to the cubes of their mean distances from the Sun. Knowledge of these laws, especially the second (the law of areas), proved crucial to Sir Isaac Newton in 1684–85, when he formulated his famous law of gravitation between Earth and the Moon and between the Sun and the planets, postulated by him to have validity for all objects anywhere in the universe. Newton showed that the motion of bodies subject to central gravitational force need not always follow the elliptical orbits specified by the first law of Kepler but can take paths defined by other, open conic curves; the motion can be in parabolic or hyperbolic orbits, depending on the total energy of the body. Thus, an object of sufficient energy—e.g., a comet—can enter the solar system and leave again without returning. From Kepler's second law, it may be observed further that the angular momentum of any planet about an axis through the Sun and perpendicular to the orbital plane is also unchanging.

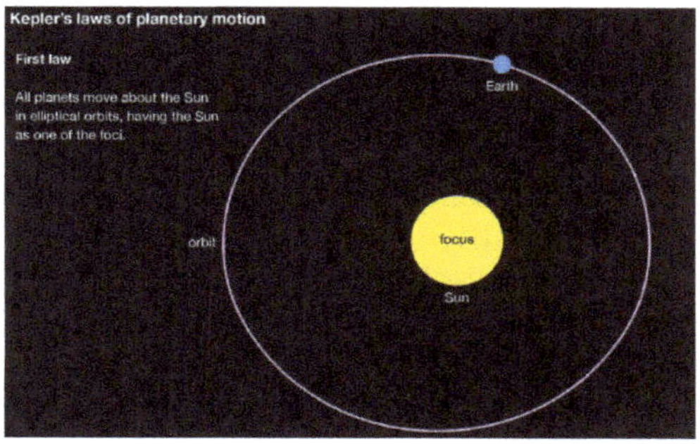

Kepler's first law: Kepler's first law of planetary motion. All planets move around the Sun in elliptical orbits, with the Sun as one focus of the ellipse.

The usefulness of Kepler's laws extends to the motions of natural and artificial satellites, as well as to stellar systems and extrasolar planets. As formulated by Kepler, the laws do not, of course, take into account the gravitational interactions (as perturbing effects) of the various planets on each other. The general problem of accurately predicting the motions of more than two bodies under their mutual attractions is quite complicated; analytical solutions of the three-body problem are unobtainable except for some special cases. It may be noted that Kepler's laws apply not only to gravitational but also to all other inverse-square-law forces and, if due allowance is made for relativistic and quantum effects, to the electromagnetic forces within the atom.

Here are the three Kepler's laws:

- The Law of Orbits: All planets move in elliptical orbits, with the sun at one focus.
- The Law of Areas: A line that connects a planet to the sun sweeps out equal areas in equal times.
- The Law of Periods: The Square of the period of any planet is proportional to the cube of the semi major axis of its orbit.

Kepler's laws were derived for orbits around the sun, but they apply to satellite orbits as well.

The Law of Orbits

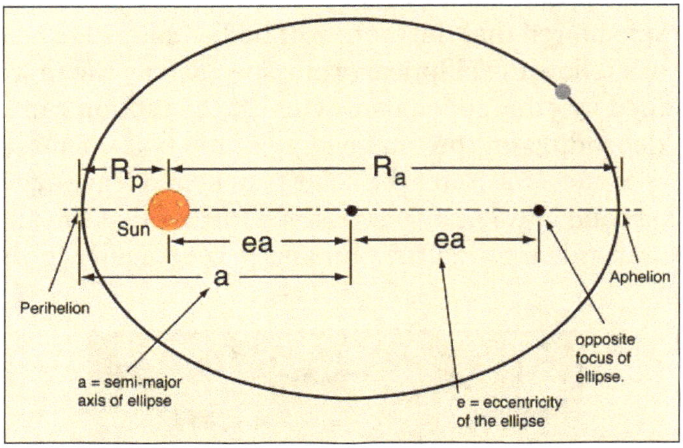

All planets move in elliptical orbits, with the sun at one focus.

$$R_a = a(1+e) \quad R_P = a(1-e)$$

This is one of Kepler's laws. The elliptical shape of the orbit is a result of the inverse square force of gravity.

Orbit Eccentricity

The eccentricity of an ellipse can be defined as the ratio of the distance,

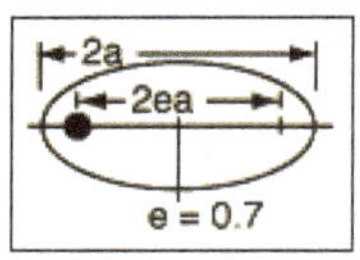

between the foci to the major axis of the ellipse. The eccentricity is zero for a circle. Of the planetary orbits, only Pluto has a large eccentricity.

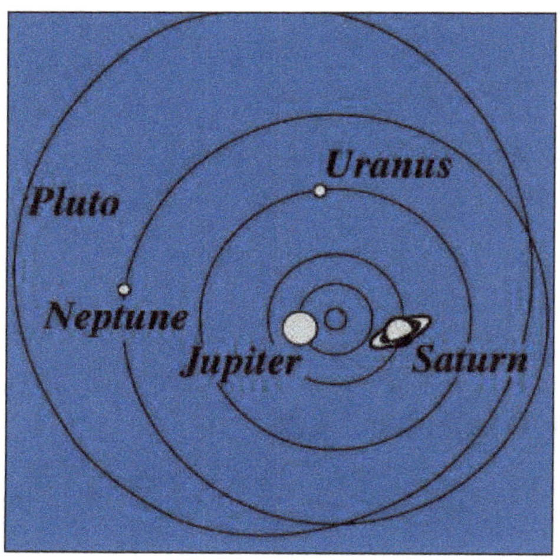

Examples of Ellipse Eccentricity

Planetary orbit eccentricities:

- Mercury .206
- Venus .0068
- Earth .0167
- Mars .0934
- Jupiter .0485
- Saturn .0556
- Uranus .0472
- Neptune .0086
- Pluto .25

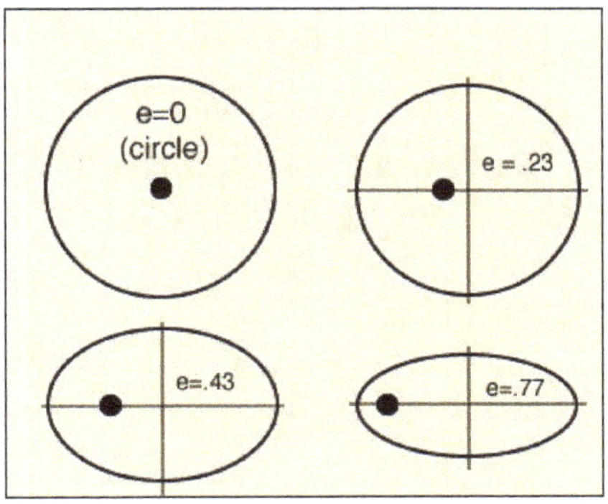

The Law of Areas

All planets move in elliptical orbits, with the sun at one focus.

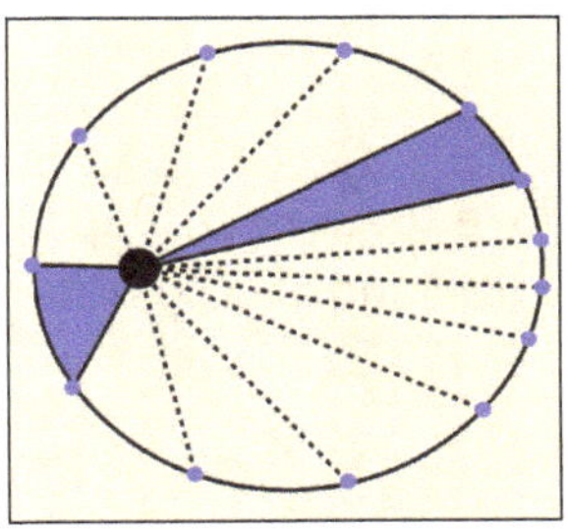

This is one of Kepler's laws. This empirical law discovered by Kepler arises from conservation of angular momentum. When the planet is closer to the sun, it moves faster, sweeping through a longer path in a given time.

The Law of Periods

The square of the period of any planet is proportional to the cube of the semimajor axis of its orbit.

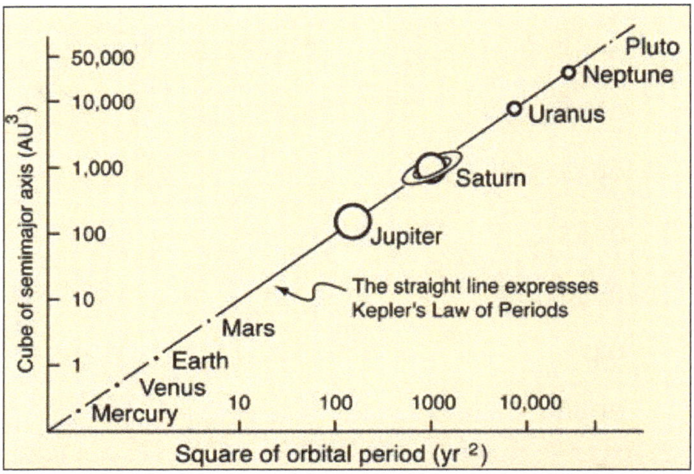

$$T^2 = \frac{4\pi^2}{GM} a^3$$

This is one of Kepler's laws. This law arises from the law of gravitation. Newton first formulated the law of gravitation from Kepler's 3rd law.

$T^2 = \frac{4\pi^2}{GM} a^3$ can be expressed simply $T^2 = a^3$

If expressed in the following units:

- T Earth years,
- a Astronomical Units AU (a = 1 AU for Earth),
- M Solar masses M⊙.

Then $\frac{4p^2}{G} = 1$

Kepler's Law of Periods in the above form is an approximation that serves well for the orbits of the planets because the Sun's mass is so dominant. But more precisely the law should be written,

$$T^2 = \frac{4\pi^2}{G(M_1 + M_2)} a^3$$

In this more rigorous form it is useful for calculation of the orbital period of moons or other binary orbits like those of binary stars.

Data: Law of Periods

Data confirming Kepler's Law of Periods comes from measurements of the motion of the planets.

Planet	Semimajor axis (10^{10}m)	Period T (y)	T^2/a^3 ($10^{-34}y^2/m^3$)
Mercury	5.79	0.241	2.99
Venus	10.8	0.615	3.00
Earth	15.0	1	2.96
Mars	22.8	1.88	2.98
Jupiter	77.8	11.9	3.01
Saturn	143	29.5	2.98
Uranus	287	84	2.98
Neptune	450	165	2.99
Pluto	590	248	2.99

The quantity T^2/a^3 depends upon the sum of the masses of the Sun and the planet, but since the mass of the Sun is so great, adding the mass of the planet makes very little difference.

The Virial Theorem

The virial theorem states that, for a stable, self-gravitating, spherical distribution of equal mass objects (stars, galaxies, etc), the total kinetic energy of the objects is equal to minus 1/2 times the total gravitational potential energy. In other words, the potential energy must equal the kinetic energy, within a factor of two.

Suppose that we have a gravitationally bound system that consists of N individual objects (stars, galaxies, globular clusters, etc.) that have the same mass m and some average velocity v. The overall system has a mass $M_{tot} = N \cdot m$ and a radius R_{tot}.

The kinetic energy of each object is K.E.(object) = $1/2\ m\ v^2$

while the kinetic energy of the total system is K.E.(system) = $1/2\ m\ N\ v^2 = 1/2\ M_{tot}\ v^2$

where v^2 is the mean of the squares of v. The gravitational potential energy of the system can be written as:

$$P.E.(system) \simeq -\frac{1}{2}G\frac{N^2m^2}{R_{toto}} = -\frac{1}{2}G\frac{M_{tot}^2}{R_{tot}}$$

We usually assume that all of the orbits travel on similar orbits that are isotropic, that is, are not flattened in any way and have no preferential direction; we say these are random orbits. The virial theorem then requires that the kinetic energy equals one half the potential energy, that is:

$$K.E. = -1/2P.E.$$

$$\frac{1}{2}M_{tot}v^2 = +\frac{1}{4}G\frac{M_{tot}^2}{R_{tot}}$$

$$M_{tot} \simeq 2\frac{R_{tot}v^2}{G}$$

Therefore, we can estimate the Virial Mass of a system if we can observe:

- The true overall extent of the system R_{tot}.
- The mean square of the velocities of the individual objects that comprise the system.

If the motions are not random/isotropic, the virial theorem still applies, but its form changes a bit. Similarly, since our system is made up of many objects, we can gain some insight by seeing how the orbital velocities vary with radius from the center outward.

For example, in a spiral galaxy, the dominant motion of the stars in the disk is circular rotation in the plane of the disk. The variation in the orbital velocities with radius V(r) is called the rotation curve.

Astrophysical Plasma

An astrophysical plasma is a plasma (an ionised gas) found in astronomy whose physical properties are studied in the science of astrophysics. The vast majority of the volume of the universe is thought to consist of plasma, a state of matter in which atoms and molecules are so hot, that they have ionized by breaking up into their constituent parts, negatively charged electrons and positively charged ions. Although influenced by gravity, because the particles are charged, they are also strongly influenced by electromagnetic forces, that is, by magnetic and electric fields.

All known astrophysical plasmas are magnetic. They also contain equal numbers of electrons and ions so that they are electrically neutral overall. And because plasmas are highly conductive, any charge imbalances are readily neutralised. However, because plasma phenonenon are very complex, charge imbalances can occur, resulting in a characteristic known as quasineutrality. An example is the influence of our Sun's magnetic field on the electrons and ions in the interplanetary medium (or Solar wind) resulting in the heliospheric current sheet, the largest structure in the Solar system.

Characteristics

Space plasma pioneers Hannes Alfvén and Carl-Gunne Fälthammar divided cosmic plasmas into three different categories (note that other characteristics of low-particle-density interstellar and intergalactic plasmas, means that they are characterised as medium density).

Table: Classification of Magnetic Cosmic Plasmas.

Characteristic	Space plasma density categories			Ideal comparison
	High density	Medium Density	Low Density	
Criterion	$\lambda << \rho$	$\lambda << \rho << l_c$	$l_c << \lambda$	$l_c << \lambda$
Examples	Stellar interior Solar photosphere	Solar chromosphere/corona Interstellar/intergalactic space Ionopshere above 70km	Magnetosphere during magnetic disturbance. Interplanetary space	Single charges in a high vacuum
Diffusion	Isotropic	Anisotropic	Anisotropic and small	No diffusion
Conductivity	Isotropic	Anisotropic	Not defined	Not defined
Electric field parallel to B in completely ionized gas	Small	Small	Any value	Any value
Particle motion in plane perpendicular to B	Almost straight path between collisions	Circle between collisions	Circle	Circle
Path of guiding centre parallel to B	Straight path between collisions	Straight path between collisions	Oscillations (eg. Between mirror points)	Oscillations (eg. Between mirror points)
Debye Distance λ_D	$\lambda_D << l$	$\lambda_D << l$	$\lambda_D << l$	$\lambda_D >> l$
Magnetohydrodynamics suitability	Yes	Approximately	No	No

Note: λ = Mean free path. ρ = Lamor radius of electron. $_{\lambda D}$= Debye length. l_c= Characteristic length.

Alfvén and Fälthammar noted that the:

> “Table shows some features of interest. The first is that most densities are to be considered as very high. Except in the close vicinity of the earth there is no analogy to high-vacuum phenomena. The laboratory analogy of cosmic space is not the vacuum in a tank but a highly ionized gas of a very high density.
>
> “Still more striking than the high densities are the very strong magnetic fields in the cosmos. In fact they are so strong that at present our laboratory resources do not suffice to produce field strengths large enough for model experiments.

"The powerful magnetic fields have two important consequences. The first is that the motion of charged particles is usually of a different type from that we are familiar with in the laboratory. The radius of curvature is very small and the particles more in the direction of the magnetic field or 'drift' perpendicular to it.

"The other consequence is that strong electric fields are easily produced by any motion seems to move with respect to the distant background. Half the diameter field of 10^6 gauss, a velocity of 3.10^5cm/sec causes an electric field of E = 10 e.s.u. – 3000 V/cm, and in a field of 11^{10} gauss the same velocity gives 30.10^6V/cm. Thus also the electric fields in the cosmos, when reduced to laboratory scale, are often very strong. "Finally, the time-scale transformation in the table is of interest. Solar flares, coronal arcs, and also the initial phase of a magnetic storm should be regarded as very short-lived phenomena. In fact their equivalent duration is of the order of the ignition time of an electric discharge. This means that transient phenomena are very important in cosmical physics".

Astrophysical plasma may be studied in a variety of ways since they emit electromagnetic radiation across a wide range of the electromagnetic spectrum. For example, cosmic plasmas in stars emits light as can be seen by gazing at the night sky. And because astrophysical plasmas are generally hot, (meaning that they are fully ionized), electrons in the plasmas are continually emitting X-rays through a process called bremsstrahlung, when electrons nearly collide with atomic nuclei. This raditation may be detected with X-ray observatories, performed in the upper atmosphere or space, such as by the Chandra X-ray Observatory satellite. Space plasmas also emit radio waves and gamma rays.

Anomalous Characteristics

In his summary of low-density plasmas, Fälthammar notes that:

"Low-density plasmas fill the rest of the universe; i.e., they dominate planetary magnetospheres, the solar corona, and interplanetary, interstellar, and intergalactic space. They represent a very complicated state of matter which can entertain a host of waves and instabilities. Their electrodynamic properties are such that in general there does not even exist a local macroscopic relation between the electric field and electric current density. For example, finite electric current may coexist with a zero electric field, and vice versa. This means that the concept of conductivity as a local property becomes meaningless, and even remote parts of the current circuit must be taken into account.

Pseudo-plasma vs. Real Plasma

First approach (pseudo-plasma)	Second approach (real plasma)
Homogeneous models	Space plasmas often have a complicated inhomogeneous structure

Conductivity $\sigma_E = \infty$	σ_E depends on current and often suddenly vanishes
Electric field $E_{\|\|}$ along magnetic field = 0	E often <> ∞
Magnetic field lines are "frozen-in" and "move" with the plasma	Frozen-in picture is often completely misleading
Electrostatic double layers are neglected	Electrostatic double layers are of decisive importance in low-density plasma
Instabilities are neglected	Many plasma configurations are unrealistic because they are unstable
Electromagnetic conditions are illustrated by magnetic field line pictures	It is equally important to draw the current lines and discuss the electric circuit
Filamentary structures and current sheets are neglected or treated inadequately	Currents produce filaments or flow in thin sheets
Maxwellian velocity distribution	Non-Maxwellian effects are often decisive Cosmic plasmas have a tendency to produce high-energy particles
Theories are mathematically elegant and very "well developed"	Theories are not very well developed and are partly phenomenological

Astrometry

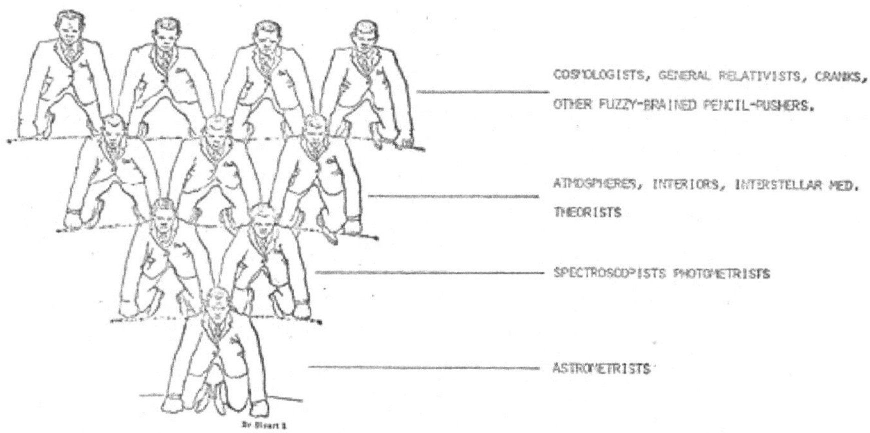

The astronomical pyramid. This image illustrates that most astronomers these days are working in extragalactic astronomy, theory, and cosmology. They all depend on observations provided by fewer astronomers. The basis is formed by the very few astrometrists.

Astrometry is the science which deals with the positions and motions of celestial objects. Astrometry is now one of many fields of research within astronomy. Historically, astrometry was all that astronomy was about until about the 19th century. Toward the end of the 19th century not only the directions, i.e. angles between celestial objects as seen on the celestial sphere were measured but also the "quality of light", specifically the light intensity (photometry) and color (spectroscopy, light intensity as function of color or wavelength). This was the birth of astrophysics. The term astrophysics, often used to distinguish most of current astronomical research from the classical astronomy (i.e. astrometry) is misleading, because astrometry also is certainly part of physics or astrophysics. Measurements of distances to celestial objects by triangulation for example is at the core of astrometry and it forms the basis of all astrophysics; without knowing the distances to planets, satellites, stars, and galaxies, no correct understanding of the cosmos in which we live can be achieved.

Angles

Astrometry is about measuring angles, dealing with errors in angular measures and changes of angles with time (angular velocity), and derivation of astrophysical quantities from those measurements. A full circle can be divided into 360 degrees. A degree is subdivided into 60 arcminutes (arcmin) and 1 arcmin equals 60 arcseconds (arcsec). The full moon in the sky substends an angle of about 1/2 degree or 30 arcmin as seen from earth. The smallest angular separations or resolutions seen through an ordinary telescope on the ground is about 1 arcsec, limited by the turbulence of earth's atmosphere. Progress in astrometry in the recent decade called for smaller angular units. A milliarcsecond (mas) is 1/1000 arcsec and a microarcsecond is 1/1000 mas. The diameter of a large coin as seen from a distance of about 6000 km (New York to London) corresponds to an angle of 1 mas.

Celestial Coordinates

Astronomers use celestial coordinates on the sky to define a position (direction unit vector) of a celestial object as seen from a specific location (which can be on earth or in space) in a way similar to geographic latitude and longitude. The most commonly used system is the equatorial celestial coordinate system which has the plane of earth's equator projected onto the celestial sphere as fundamental plane. Right ascension (RA) is the angle counted in this plane from 0 to 24 hour, similar to the geographic longitude. Declination (Dec) is the angle orthogonal to RA, i.e. the angular distance from the equatorial plane with +90 degree for the celestial north pole and -90 degree for the celestial south pole. An object on the celestial equator has Dec = 0. A more detailed narrative with figures about celestial coordinate systems is given by G.Kaplan.

The celestial coordinate system is established by large-angle, fundamental observations. These types of observations allow us to define a coordinate system (directions of 3 orthogonal axes) from first principles, without prior knowledge of the coordinates of stars. In the past these fundamental observations were provided by transit circle (meridian circle) telescopes at optical wavelengths for about 1500 bright stars. These observations were tied into the complex motion and rotation of the earth as well as other solar system bodies (sun, major and minor planets) to be able to establish a dynamical reference frame, which is made inertial (i.e. rotation free) based on celestial mechanics and law of gravitation.

In 1997 the International Astronomical Union (IAU) adopted the International Celestial Reference System (ICRS) based on more precise Very Long Baseline Interferometry (VLBI) observations with radio telescopes to define the axes of the celestial coordinate system. The catalog of about 600 compact, extragalactic sources (mainly quasars) form the International Celestial Reference Frame (ICRF), the practical realization of the ICRS. The definition of the ICRS thus is independent of the earth's motion, rotation, and its equator and is no longer a dynamical system, rather a quasi-inertial reference system with the assumption that those quasars do not move noticeably along the sky due to their enormous distances and thus are used as fixed, fiducial points to define the ICRS. The positions of most sources in the ICRF are known to about 0.1 to 1 mas, and there are no angular motions by definition.

Between 1989 and 1993 the European space mission Hipparcos observed (at visible light wavelengths) about 118,000 stars at a mean epoch of 1991.25 to an accuracy of about 1 mas per position coordinate. Because stars move, the Hipparcos Catalog also had to solve for the effects of proper motion and parallax as well as just position. The Hipparcos Celestial Reference System (HCRS) was adopted by the IAU as the optical realization of the ICRS. The coordinate system of Hipparcos was aligned to the fundamental ICRS with a variety of methods with 12 radio stars (visible at optical and radio wavelengths) providing the strongest link.

Stars do Move

Contrary to planets the stars seem to be at fixed positions on the sky and the familiar constellations don't change over years and centuries. However, if one looks more closely, all stars, including our Sun, do move in more or less regular orbits around the center of our Milky Way Galaxy. The projection of the real space motion of a celestial object (with respect to our solar system) onto the celestial sphere is called proper motion. This is an angular velocity (angle per time). The corresponding velocity (speed) is the tangential velocity. When the distance to an object is known, the tangential velocity can be calculated from its angular velocity. The third component of the 3-dimensional space motion of a celestial object is the motion along the line of sight (toward or away from us), which is called the radial velocity. It is mainly measured by spectroscopic methods and recently also became a topic of astrometry. High accuracy angular measures can detect changes in proper motions due to radial velocity and in 2000 the IAU adopted the resolution C2 defining "astrometric radial velocity".

There is another important motion of the stars: the apparent, annual parallactic motion. A nearby star seems to move along a small elliptical pattern with respect to distant stars over the course of a year due to the motion of the earth around the sun. You can visualize this effect by holding up a finger in front of you and looking at it alternatively with one and then the other eye. The finger seems to move with respect to the distant background. Half the diameter of this parallactic motion is called the parallax, p of a star. It is related to the distance, d of the star by: $d = 1 / p$ where p is measured in arcsec and d in parsec, the standard distance unit in galactic and extragalactic astronomy. A parsec is about 3.26 light years (ly), and 1 ly = 9.467 * 10^{15} meter. A parsec is thus the distance at which the mean distance between earth and sun extends an angle of 1 arcsec. The closest star beyond our sun belongs to the alpha centauri system with d = 1.3 pc and p = 0.77 arcsec. All other stars have smaller parallaxes.

Proper motions and parallaxes were typically measured over long periods of time with large (high magnification or large plate scale) telescopes on photographic plates. This is the domain of narrow-field, small-angle, differential astrometry. However, the Hipparcos satellite and other future projects are capable of measuring small, local position changes like parallax and proper motions from large angle, global, absolute, observations. On the other hand, proper motions for most of the millions of faint stars are still known today from differential comparisons of pairs of photographic plates taken of the same area of the sky decades apart. An excellent account on star catalogs up to about 1970 is given in Eichhorn. Differential observations measure the positions of celestial objects in a small area of the sky relative to other objects. This is technically easier than absolute, large-angle measures and often yields higher precision than the wide-angle measures.

Another important area of astrometry is the research of double and multiple stars. Astrometric observations and determination of orbits is the only way to directly measure the masses of stars. The mass of a star is the fundamental quantity which determines the evolution and appearance of a star throughout its life. Orbital motion of course complicates the determination of parallax and proper motions, and double stars are sometimes called the "vermon" of the sky by astrometrists not working in the field of double stars. Fortunately (from the point of view of deriving masses) and unfortunately (from the point of view of using "clean" fiducial points of light for astrometric reference frame work) most stars belong to a double or multiple star system.

Finally, the era of detecting exoplanets, i.e. planets outside our solar system which orbit around other stars by astrometric means has just begun. Astrometric "wobbles" will be observed by space missions which are already being constructed. This is similar to detecting orbital motions of invisible components of certain double stars, which for example lead to the discovery of white dwarf stars in the 19th century. The astrometric method is the only way to determine the masses of those components which could either be stars or planets except when a pair of objects is seen transiting each other (orbital plane is along our line of sight).

Hydromagnetic Dynamo Theory

Dynamo theory is a vast field with almost a hundred years of history, starting with early ideas by Joseph Larmor in 1919. Dynamos are particularly important in connection with understanding magnetic fields in astrophysics. Much of the work is at the level of analytic theories and numerical simulations. During the last decade also various liquid metal experiments have been performed.

Traditionally, dynamos are divided into:

- Kinematic dynamos, where the flow can be considered given, and
- Nonlinear dynamos, where the flow is affected by the magnetic field through the Lorentz force.

The latter are sometimes also referred to as hydromagnetic dynamos, which emphasizes the importance of hydromagnetic interactions.

Kinematic Dynamos

For kinematic dynamos the field strength is negligible, so the flow can be considered given. It must still be a solution to the Navier-Stokes equations, although this restriction is sometimes ignored. With given boundary conditions (e.g. vacuum outside the dynamo domain) this constitutes an eigenvalue problem, where the largest real part of the eigenvalue is the growth rate of the magnetic field. The dynamo is excited if the growth rate is positive.

Kinematic dynamos may be divided into laminar and turbulent dynamos. For laminar dynamos the velocity field is spatially smooth and often stationary. For these dynamos there is a further distinction between slow dynamos and fast dynamos, depending on whether in the limit of vanishing magnetic resistivity the growth rate of the magnetic field goes to zero (slow dynamo) or remains finite (fast dynamo). Examples of slow dynamos include the Roberts flow, which is a two-dimensional helical flow pattern, as it is realized approximately in the Karlsruhe dynamo experiment. An example of a fast dynamo is the ABC flow, although here the evidence is only numerical.

Turbulent dynamos can be divided into small-scale and large-scale dynamos. Small-scale dynamos work in principle with any random flow, but here we restrict ourselves to the physically relevant case of fully isotropic turbulence. Such flows constitute an idealization that allows analytic progress to be made. Such analytic theories can also be tested numerically, although the maximum mesh sizes restrict the largest achievable Reynolds numbers currently to around 1000. Real flows are always to some degree anisotropic due to the nature of the underlying forcing mechanism.

The onset of dynamo action is characterized by the magnetic Reynolds number, which is defined as $R_m=u_{rms}/(\eta k_f)$. Here, u_{rms} is the root-mean-square velocity in the dynamo-active domain, η is the magnetic diffusivity, and k_f is the wavenumber corresponding to the energy-carrying scale of the flow.

Small-scale Dynamos

Small-scale dynamos can be studied analytically by solving evolution equations either for the correlation function $<B_iB_j>$ or for the energy spectrum E(k).

The critical magnetic Reynolds number for the onset of dynamo action is around R_m=35. The magnetic field has an approximate $k^{3/2}$ energy spectrum and is peaked at the resistive scale ~η/u_{rms}. The growth rate scales with $Rm^{1/2}$.

At low magnetic Prandtl numbers $P_m=\nu/\eta$, the dynamo becomes harder to excite. This result does not, however, apply to dynamos with a mean flow (for example the Taylor-Green flow has a finite time average), or to flows with finite net helicity or anisotropy.

Large-scale Dynamos

For large-scale dynamos the magnetic energy grows at scales large compared with the scale of the turbulence. This requires that there is scale separation, i.e. that the domain size is large compared with the size of the turbulent eddies.

The evolution equation for the large-scale field is obtained by averaging the Faraday equation (or induction equation; i.e.

$$\frac{\partial \overline{B}}{\partial t} = \nabla \times \left(\overline{\mathrm{U}} \times \overline{\mathrm{B}} + \overline{\mathrm{u} \times \mathrm{b}} - \eta \mu_0 \overline{\mathrm{J}} \right),$$

where the expression $\overline{u} \times \overline{b} = \overline{\varepsilon}$ is also referred to as the mean turbulent electromotive force. In the simplest isotropic case one has,

$$\overline{\varepsilon} = \alpha \overline{B} - \eta_t \mu_0 \overline{\mathrm{J}}$$

Here the first term is called the α; effect, while the second term corresponds to turbulent diffusion with a turbulent diffusivity, η_t. In the anisotropic case there can be many more terms. One particularly important one that can also lead to dynamo action is the so-called $\overline{W} \times \overline{\mathrm{J}}$ effect.

Large-scale dynamos are amenable to a mean field treatment where one considers only the averaged equations using, for example, an azimuthal average relevant for axisymmetric fields. In analytic studies one often uses instead ensemble averages.

Nonlinear Dynamos

The nonlinear regime is reached when the Lorentz force begins to affect the fluid motions. When a kinematic dynamo has achieved appreciable amplitudes at the end of its exponential growth, the Lorentz force will usually begin to quench the dynamo and lead to some equilibration. This is the normal situation. There are, however, two rare exceptions: there can be so-called self-driving dynamos (where a suitable flow only exists because of the Lorentz force) and so-called self-killing or suicidal dynamos (where the Lorentz force destabilizes the flow that led to the exponential growth in the kinematic regime). An important example of a self-driving dynamo is the dynamo that works as a result of the magnetorotational instability. Another example are the numerical models of the geodynamo.

Saturation of Small-scale Dynamos

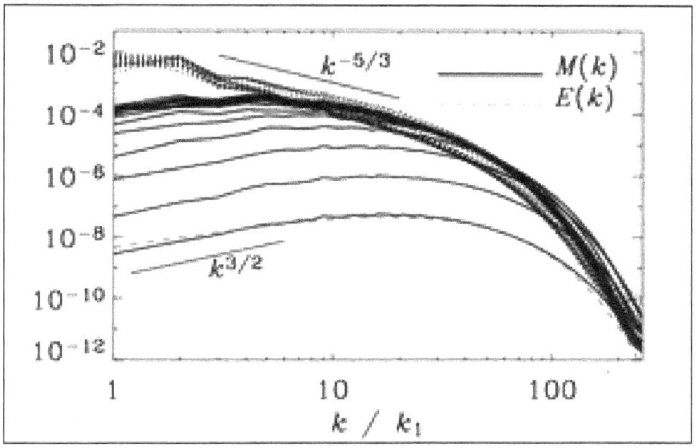

Magnetic and kinetic energy spectra from a nonhelical turbulence simulation with P_m=1.

The kinetic energy is indicated as a dashed line (except for the first time displayed where it is shown as a thin solid line). At early times the magnetic energy spectrum follows the $k^{3/2}$ Kazantsev law, while the kinetic energy shows a short $k^{-5/3}$ range.

In the kinematic regime the magnetic energy spectrum develops a $k^{3/2}$ power law at large scales, so the spectral magnetic energy peaks at small scales. As the dynamo saturates, the magnetic energy spectrum approaches the $k^{-5/3}$ power law of the turbulent flow and saturates. Thus, also the magnetic energy spectrum gradually develops a $k^{-5/3}$ power law that is familiar from Kolmogorov turbulence. Simulations can at present only show the beginnings of this development.

Saturation of Large-scale Dynamos

What we said about the saturation of small-scale dynamos does in principle also apply to large-scale dynamos. However, in certain cases large-scale dynamos can saturate slowly or at substantially lower field strengths due to the effects of magnetic helicity conservation. This applies in particular to the case of closed or periodic boundary conditions that are often used in numerical simulations. Writing the induction equation $\partial B/\partial t = -\nabla \times E$ in terms of the magnetic vector potential A, where $B = \nabla \times A$, we have $\partial A/\partial t = -E - \nabla\varphi$, where φ is the electrostatic potential. This way we obtain an evolution equation for the magnetic helicity <A·B> of the form,

$$\frac{\partial}{\partial t}\langle A \cdot B\rangle = -2\eta\mu_0 \langle \mathrm{J} \cdot B\rangle,$$

Using the fact that $E = -U \times B + J/\sigma$, where $\sigma = 1/(\eta\mu_0)$ is the electric conductivity, we see that,

$$\frac{\partial}{\partial t}\langle A \cdot B\rangle = -2\eta\mu_0 \langle \mathrm{J} \cdot B\rangle,$$

so the magnetic helicity can only evolve resistively. This is what slows down the saturation of those large-scale dynamos where magnetic helicity plays a role and what can thus lead to catastrophically low saturation levels.

The evolution equation for the magnetic helicity of the large scale field is similar to that for the total field, except that the mean electromotive force produces additional magnetic helicity proportional to,

$$\overline{\varepsilon}\cdot\overline{B}=\alpha\left\langle\overline{B}^2\right\rangle-\eta_t\mu_0\left\langle\overline{J}\cdot\overline{B}\right\rangle.$$

The evolution equation for the magnetic helicity in the small scale field, <a·b>, must then be amended by the same term, but with the opposite sign so that the sum of both terms adds up to the original helicity equation. This leads to the production of excess small scale magnetic helicity which, in turn, modifies the total (magnetic and kinetic) small scale helicity and thus quenches the α effect.

Astrophysical bodies escape this type of quenching by developing magnetic helicity fluxes inside the dynamo domain and out through the boundaries. In the case of the sun the shedding of magnetic helicity may occur mostly via coronal mass ejections.

Applications

Sun

Longitudinally averaged radial component of the observed solar magnetic field as a function of cos(colatitude) and time. Dark (blue) shades denote negative values and light (yellow) shades denote positive values. Note the sign changes both in time and across the equator.

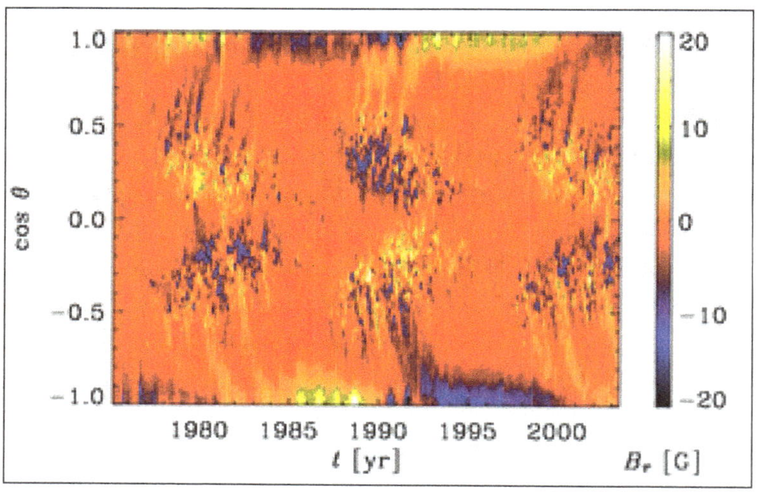

The sun has a magnetic field that manifests itself in sunspots through Zeeman splitting of spectral lines. It has long been known that the sunspot number varies cyclically with a period between 7 and 17 years. The longitudinally averaged component of the radial magnetic field of the sun shows a markedly regular spatio-temporal pattern where the radial magnetic field alternates in time over the 11 year cycle and also changes sign across the equator. One can also see indications of a migration of the field from mid-latitudes toward the equator and the poles. This migration is well seen in a sunspot diagram, which is also called a butterfly diagram, because the pattern formed by the positions of sunspots in time and latitude looks like a sequence of butterflies lined up along the equator.

Distributed vs. Overshoot Dynamos

There is at present no clear consensus as to whether the solar dynamo operates in the entire convection zone of the sun (distributed dynamo) or whether it works preferentially at the bottom of the convection zone in the so-called tachocline, where the differential rotation changes sharply into a rigidly rotating profile.

Advection Dominated Dynamos

In recent years the idea of advection-dominated or flux transport dynamos has been developed. Here the cycle period and the propagation direction of the magnetic activity pattern is determined by a large scale meridional circulation. There is now observational evidence that such a circulation of suitable magnitude and direction does indeed exist.

Galaxies

Observations of radio emission of spiral galaxies show well developed large scale patterns. The flows responsible for dynamo action in galaxies include both the differential rotation and supernova-driven turbulence. The e-folding time for dynamo action may be a significant fraction of the age of the universe, so the initial seed magnetic fields must not be too weak. As possible sources of seed magnetic fields outflows from active galaxies and starburst galaxies, primordial magnetic fields, and battery effects are being discussed. Primordial magnetic fields may have been generated during one of the phase transitions during the early Universe.

Accretion Discs

The rotation law in accretion discs is Keplerian, resulting from a balance between centrifugal and gravitational forces ($v\varphi^2/r=GM/r^2$), where v_φ is the azimuthal velocity, r is the radius from the central object, G is Newton's gravitational constant, and M is the mass of the central object.) Such a rotation law implies that the specific angular momentum increases with radius, so such discs are hydrodynamically stable, but in the presence of a magnetic field, points that are separated in space may be coupled nonlocally. Under such conditions two points in a Keplerian orbit that are being pulled together will actually move further apart from each other. This is the essence of the magnetorotational instability (MRI).

In practice, because of large Reynolds numbers, the MRI leads to turbulence. This turbulence, in turn, can lead to dynamo action. Simulations in the presence of stratification have shown that there can be an α effect, although the sign is opposite to the one naively to be expected.

Early Universe

There are several mechanisms that could potentially generate magnetic fields during one of the early phase transitions (e.g. the electroweak phase transition). In the absence of any additional driving, the flows would be primarily a consequence of the driving from the Lorentz force. Since that time, and before gravitational clumping takes over, the magnetic field would only be decaying. The field might however be helical, in which case a magnetic cascade would be possible that would transfer magnetic energy to larger scales.

Laboratory Plasma Dynamos

Dynamo effects are being discussed in connection with various plasma experiments. These are usually relatively short-lived events (nanoseconds to microseconds) that are initiated by an electric discharge. Prime examples are the Reversed Field Pinch and Spheromak configurations, where velocity and magnetic field fluctuations generated from a current-driven instability (current parallel to the magnetic field) correlate to produce a turbulent electromotive force. This generates a poloidal field via an analogous α effect. Here, unlike in the case of kinematic velocity driven dynamos, the α effect is driven by an externally imposed magnetic field.

Laboratory Liquid Metal Dynamos

Over the past few years it has been possible to demonstrate sustained self-excited dynamo action in the laboratory using liquid sodium. Although the magnetic Reynolds numbers are currently lower than what is possible with simulations, the fluid Reynolds numbers are much higher than what can be achieved in simulations because of the very small magnetic Prandtl number. In this respect we may expect a lot of new results to come from laboratory experiments.

Astrophysical Magnetohydrodynamics

In a highly collisional plasma with perfect conductivity, the equations of motion are essentially the Euler equations of gas dynamics, supplemented with Maxwell's equations to describe the evolution of the magnetic field (in particular, Faraday's Law). The result, usually referred to as the equations of ideal MHD, is

$$\frac{\partial}{\partial t}+\nabla.[p\mathrm{v}]=0,$$
$$\frac{\partial p\mathrm{v}}{\partial t}+\nabla.\left[p\mathrm{vv}-\mathrm{BB}+\mathrm{P}^{*}\right]=0,$$
$$\frac{\partial E}{\partial t}+\nabla.\left[\left(E+P^{*}\right)\mathrm{v}-\mathrm{B}(B.\mathrm{v})\right]=0,$$
$$\frac{\partial B}{\partial t}-\nabla\times(\mathrm{v}\times B)=0,$$

where P* is a diagonal tensor with components $P^* = P + B^2/2$ (with P the gas pressure), E is the total energy density,

$$E=\frac{P}{\gamma-1}+\frac{1}{2}p\mathrm{v}^{2}+\frac{B^{2}}{2},$$

and $B^2 = B \cdot B$. The other symbols have their usual meaning. These equations are written in units such that the magnetic permeability μ = 1. An equation of state appropriate to an ideal gas, P = (γ − 1)e (where γ is the ratio of specific heats, and e is the internal energy density), has been assumed in writing the above equation. These equations are valid only for non-relativistic flows, and for phenomena at frequencies much less than the plasma frequency. There are many interesting frontiers to explore as some of the assumptions underlying the equations of ideal MHD are relaxed, for example in low collisionality plasmas.

Restricting ourselves to one dimensional flow for the moment, it is useful to rewrite the equations of motion in a compact form,

$$\frac{\partial U}{\partial t}=\frac{\partial F}{\partial U}\frac{\partial U}{\partial x}$$

where the components of the vectors U and F are the conserved variables and their fluxes, respectively, that is,

$$U=\begin{bmatrix} p \\ M_x \\ M_y \\ M_z \\ E \\ B_y \\ B_z \end{bmatrix},\quad F=\begin{bmatrix} pv_x \\ pv_x^2+P+B^2/2\ -B_x^2 \\ pv_xv_y-B_xB_y \\ pv_xv_z-B_xB_z \\ \left(E+P^{*}\right)v_x-(B\cdot v)B_x \\ B_yv_x=B_xv_y \\ B_zv_x-B_xv_z \end{bmatrix},$$

Equation $\frac{\partial U}{\partial t}=\frac{\partial F}{\partial U}\frac{\partial U}{\partial x}$ defines a system of nonlinear hyperbolic partial differential equations (PDEs). The mathematical properties of hyperbolic PDEs are well studied.

In particular, the eigenvalues of the Jacobian ∂F/∂U define the characteristic (wave) speeds in MHD.

In fact, probably the most important property of hyperbolic PDEs is that they admit wave-like solutions. Much of the dynamics of magnetized plasmas can be interpreted using the properties of linear and nonlinear wave solutions. The properties of linear waves can be studied using the dispersion relation, derived by looking for solutions for small amplitude disturbances of the form exp i(ωt + k · x), where ω is the frequency and k the wavevector, in a stationary, isotropic, homogeneous medium. Inserting this form for the solution into the equations of motion, and keeping only terms which are linear in the disturbance amplitude results in a system of linear equations, which have solutions only if the frequency and wavenumber are related through the following dispersion relation,

$$\left[\omega^2-(k\cdot V_A)^2\right]\left[\omega^4-\omega^2k^2\left(V_A^2+C^2\right)+K^2C^2\left(K\cdot V_A\right)^2\right]=0$$

where $V_A = B/\sqrt{4\pi p}$ is the Alfven velocity, and $C^2 = \gamma P/\rho$ the adiabatic sound speed. The dispersion relation has three pairs of solutions, which represent right- and left-going waves of three different families. The MHD wave families are the Alfven wave (an incompressible transverse wave propagating at speed VA), and the fast and slow magnetosonic waves (which are both compressible acoustic modes with phase velocity modified by the magnetic pressure). To complicate matters even more, the phase velocity for each mode depends on the angle between the wavevector and the magnetic field, as well as the strength of the magnetic field as measured by the ratio VA/C. The angular dependence is most easily demonstrated using Friedrichs diagrams, which plot the relative phase velocity of each mode versus the angle between k and B in a polar diagram. Such plots clearly demonstrate important properties of MHD waves, for example for directions parallel to the magnetic field, the Alfven wave has the same phase velocity as either the fast or the slow´ magnetosonic wave (which one depends on whether the Alfven speed is faster or slower than the´ sound speed). In this case, the modes are degenerate. Mathematically, this reflects the fact that the equations of MHD are not strictly hyperbolic, since in some circumstances the eigenvalues of the Jacobian are degenerate. This fact makes finding solutions to the equations of ideal MHD even more complicated.

Another important property of MHD waves in comparison to hydrodynamics is that, because they involve transverse motions, Alfven waves can be polarized. The sum of two linear polar-´ izations with different phase shifts can lead to circularly polarized Alfven waves. This means in ´MHD, all three components of velocity must be kept, even in one dimensional flows, in order to represent all polarizations. Moreover, in non-ideal MHD the left- and right-circularly polarized Alfven waves can have different phase velocities, these are the whistler waves in the Hall MHD regime (where ions and electrons can drift due to collisions with neutrals). Again, this new behavior is a direct consequence of the complexity of MHD waves, and it is fair to say that the rich dynamics of MHD results in part from this complexity.

Finding analytic solutions to the equations of MHD, beyond those representing linear waves, is very difficult. Usually, very restrictive assumptions are required, such as steady (so that $\partial/\partial t = 0$), and/or one dimensional flow. Today, the most important tools for solving the MHD equations are numerical methods. Grid based methods for MHD are now quite mature, and a variety of public codes are available to study MHD flows in fully three-dimensions, including a rich set of physics beyond ideal MHD. Most grid based methods for MHD adopt the same approach: the conserved variables are discretized on a grid, with volume averaged values stored at cell centers. In order to enforce the divergence-free constraint on the magnetic field, it is better to store area averages of each component of the magnetic field at corresponding cell faces, and evolve these components using electric fields at cell edges, using a technique called "constrained transport".

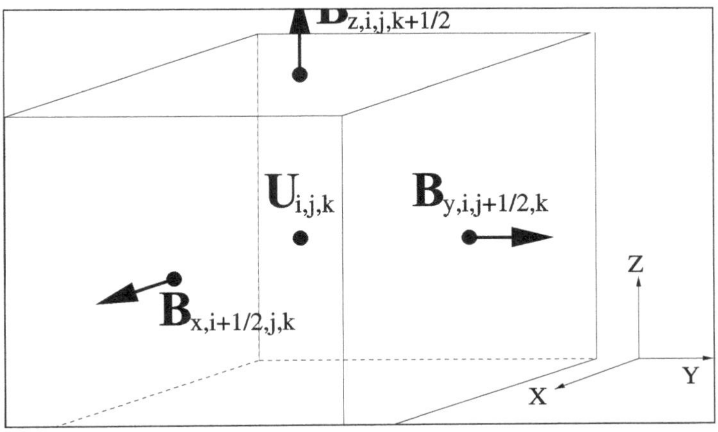

Basic centering of variables for a grid-based numerical method for MHD using constrained transport. Volume averages of conserved variables are stored at cell centers, while area averages of each component of the magnetic field are stored at cell faces.

Athena implements a higher-order Godunov scheme based on directionally unsplit integrators, piecewise-parabolic reconstruction, and constrained transport, with a variety of Riemann solvers available to compute the fluxes. With this approach, mass, momentum, energy, and magnetic flux are all conserved to machine precision. Of course, there are many other codes available which implement different algorithms than those used in Athena, and this is a very good thing, because by comparing solutions to the same problem generated by different algorithms, we can gauge whether those solutions are reliable.

Solar Magnetoconvection

The best evidence of the importance of magnetic fields to the dynamics of astrophysical plasmas comes from observations of the outer layers of the Sun. Both the presence of sunspots in the photosphere, and structures such as filaments, prominences, and flares in the solar corona, demonstrate the key role that magnetic fields play in shaping the dynamics. In fact, the very existence of the hot corona is now interpreted as due to heating by MHD effects. Beautiful images and animations that show magnetic fields in action in the solar corona have been obtained by recent spacecraft missions such as SOHO, TRACE, Yokoh, Hinode, and SDO.

It is thought that most of the magnetic activity of the Sun is driven by the combination of rotation and turbulent flows in the convection zone. In fact, the properties of MHD turbulence driven by convection was one of the problems that first interested Chandra in plasma physics.

Understanding the origin and evolution of the Sun's magnetic field via a dynamo process has been a challenging problem for many decades. In addition to generation of the dipole field due to differential rotation, a process first proposed by Parker, there are also small-scale multiple fields thought to be generated by the convective turbulence that play a role in shaping sunspots and coronal activity. Both the processes that produce sunspots, and the large-scale magnetic field of the Sun, are very active areas of research.

In the case of sunspots, direct numerical simulations of magnetoconvection in the outer layers, including realistic radiative transfer to capture the outer radiative zone, can now reproduce details of observed sunspots, including the penumbral filaments.

In the case of the solar dynamo, the dipole field is now thought to originate in the tachocline, a region of strong shear between the radiative core (which is in solid body rotation, according to results from helioseismology) and the outer convective zone (which is in differential rotation). However, although the sophistication of modern global MHD simulations of magnetoconvection in spherical and rotating stars is impressive, they still fail to explain both the origin of the differential rotation in the convective zone, and the origin of the cyclic dipole field. Solving the solar dynamo problem is important, as we are unlikely to understand magnetic fields in other stars if we cannot first understand the Sun.

The MRI in Accretion Disks

Moving beyond stars, the next set of astrophysical systems where magnetic fields have been identified as being important is accretion disks. Such disks are ubiquitous, occurring in protostellar systems, close binaries undergoing mass transfer, and in active galactic nuclei.

The most basic property of an accretion disk is the angular momentum transport mechanism. This mechanism controls the rate of accretion, which in turn controls the luminosity, variability, and spectrum of the disk. Mass accretion in disks is analogous to nuclear fusion in stars: it is the mechanism that powers the entire system.

It has been known for decades that kinetic viscosity in an astrophysical plasma is too small to explain the angular momentum transport and mass accretion rate, so that some form of "anomalous" viscosity is required. It has also been long suspected that the transport was associated with turbulence in the disk, but disks with Keplerian rotation profiles are linearly stable according to the Rayleigh criterion, that is, so long as the specific angular momentum increases outwards. So the question becomes: what drives turbulence in disks?

The answer seems to be: magnetic fields. Remarkably, disks with Keplerian rotation profiles which contain weak magnetic fields (weak in the sense that the gas pressure is larger than the magnetic pressure) are linearly unstable to the magnetorotational instability (MRI), as first recognized by Balbus & Hawley. The MRI can be identified by calculating the linear dispersion relation for MHD waves in a Keplerian shear flow. The simplest analysis which captures the MRI assumes incompressible axisymmetric perturbations, a purely vertical magnetic field, and ideal MHD (all of these assumptions have been relaxed in later analyses. The resulting dispersion relation is,

$$\omega^4 - \omega^2\left[k^2 + 2(k\cdot V_A)^2\right] + (k\cdot V_A)^2\left([k\cdot V_A]^2 + \frac{d\Omega^2}{d\,Inr}\right) = 0$$

where V_A is the Alfven speed, and

$$k^2 = \frac{1}{R^3}\frac{d\left(R^4\Omega^2\right)}{dR}$$

is the epicyclic frequency (R is the cylindrical radius). Note that the coefficient of the first and second terms in equation $\omega^4 - \omega^2 \ldots\left([k\cdot V_A]^2 + \frac{d\Omega^2}{d\,Inr}\right) = 0$ are positive and negative respectively, therefore solutions with $\omega^2 < 0$ (that is, instability) are possible if the third term is negative. This occurs when $(k\cdot V_A)^2 < -\frac{d\Omega^2}{d\,Inr}$.

Physically, this states that if the rotation frequency in the disk is decreasing outwards (as is true in Keplerian flows), then there are always sufficiently small wavenumbers that will be unstable. Note that this is in direct contradiction to the Rayleigh criterion, which requires the angular momentum (not frequency) decrease outward for instability. How small is "sufficiently small" for instability depends on the magnetic field strength (VA). In practice, if the field is weak (VA < C), there always are unstable modes with wavenumbers large enough that the corresponding wavelength is less than the vertical scale height (thickness) of the disk.

In fact, studies of the MRI have a long and interesting history. The MRI was first identified by Velikhov in a study motivated by a rotating plasma experiment. Chandrasekhar made important contributions, showing the instability was present in a global analysis of magnetized Couette flow. Fricke found the instability in differentially rotating stars. However, the importance of the MRI to accretion disks was not recognized by any of these authors, in fact Safronov argued that the inclusion of finite resistivity and viscosity effects would make the MRI unimportant in disks. A key element of confusion seems to be over the lack of recovery of the Rayleigh criterion as the magnetic field strength is decreased to zero. The stability properties of hydrodynamic flows (based on angular momentum gradients) and MHD flows (based on angular velocity gradients) are incompatible, a point discussed in detail by Balbus & Hawley.

It was not until their paper that the important role that the MRI plays in disks was identified. Over the past 20 years, there has been considerable effort to understand the nonlinear regime and saturation of the MRI, mostly using computational methods. Figure shows images from typical simulations of the MRI in both global domains, in which the entire disk is evolved over a wide range of radii, and local shearing box simulations, in which only a small radial extent of the disk is evolved. The advantage of the shearing box is that by focusing all of the computational resources on a small patch, much higher numerical resolution is possible.

Perhaps the most important result from local shearing box simulations is that in the nonlinear regime, the MRI produces MHD turbulence which has both significant Maxwell and Reynolds stresses that transport angular momentum outward. It is remarkable that the inclusion of a weak field qualitatively changes the stability properties of the flow, and results in outward transport at a level required by observations. Numerical simulations of the MRI have also established that turbulence amplifies the magnetic field, and drives an MHD dynamo, and that the power spectrum of the turbulence is anisotropic, with most of the energy on the largest scales.

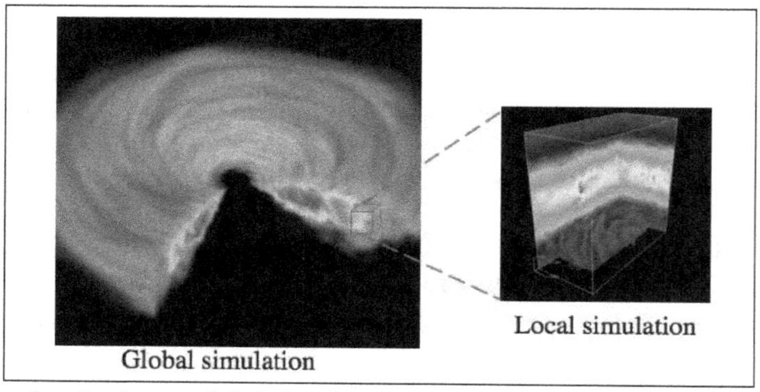

Images of the density from a global simulation of a MRI unstable disk (left), and of the density and magnetic field vectors from a local shearing box simulation (right).

Still, many important questions remain. At the moment, it is not understood how the energy liberated by accretion is dissipated by the turbulence: does most of the energy go into the ions or electrons? It is not understood how MRI unstable disks drive powerful winds and outflows as are observed in many astrophysical systems, and what are the relative contributions of the MRI and winds to angular momentum transport. Finally, calculations which include radiation have only begun to be explored; it is likely many important phenomena may be related to the interaction of the radiation field with the flow field generated by the MRI. All of these questions will undoubtedly be addressed by future efforts.

MHD Turbulence in the ISM of Galaxies

Moving to ever larger scales, the next system in which magnetic fields have been observed to be important is the interstellar medium (ISM) of galaxies. The observation of polarized synchrotron emission from the ISM of the Milky Way and other galaxies, produced by relativistic electrons spiraling around magnetic field lines, is direct proof of the presence of such fields. Moreover, the observations allow the strength and even the direction of the field to be inferred. In most cases, it is found the fields are in equipartition, with the magnetic energy density being about equal to the thermal energy of the gas, and kinetic energy of relativistic particles. Moreover, observations of the kinematics of the ISM in galaxies reveal it is highly turbulent. Thus, interpretation of the dynamics of the ISM requires an understanding of highly compressible MHD turbulence.

In fact, the statistical properties of turbulence were of considerable interest to Chandra.

> "We cannot construct a rational physical theory without an adequate base of physical knowledge. It would therefore seem to me that we cannot expect to incorporate the concept of turbulence in astrophysical theories in any essential manner without a basic physical theory of the phenomenon of turbulence itself."

Fortunately, the theory of energy cascades in strong MHD turbulence has progressed enormously in the last few decades, so that there now are theories of the power spectrum and statistical properties of MHD turbulence that can be tested and compared to observation. One method to investigate the properties of MHD turbulence is through direct numerical simulation.

Figure shows images from high resolution (1024^3) numerical simulations of highly compressible MHD turbulence with both strong and weak magnetic fields, taken from Lemaster & Stone. The turbulence is driven with a forcing function whose spatial power spectrum is highly peaked at a wavenumber corresponding to about 1/8 the size of the computational domain. The energy input rate of the driving is held constant, and the turbulence is driven so that the Mach number of RMS velocity fluctuations $M = \sigma v/C$ (where C is the sound speed) is about 7. The magnetic field strength corresponds to a ratio of gas to magnetic pressure $\beta = 8\pi P/B^2$ of 0.01 in the strong field case, and one in the weak field case. This means the Alfvenic Mach number of the turbulence is about one in the strong field case, and 7 in the weak field case.

It is quite clear from the images that in the weak field case, the density fluctuations are isotropic, and the magnetic field is highly tangled. In contrast, in the strong field case the density fluctuations are elongated along the field lines, and the field is more or less ordered. This suggests that the power spectrum of the turbulence will be anisotropic. In fact, this is one of the most basic predictions of the theory.

In addition to investigating the spectrum of fluctuations, such simulations can be used to measure properties such as the decay rate of the turbulence, and how it depends on the magnetic field strength. Early predictions suggested the decay rate of strongly magnetized turbulence would be very low, since it would be dominated by incompressible Alfven waves. In fact, the simulations found the decay rate of supersonic MHD turbulence was very fast, with the decay time about equal to an eddy turn over time on the largest scales, regardless of the field strength. Most of the dissipation was found to occur in shocks. Thus, while Alfven waves are important to the energetics, the coupling of large amplitude nonlinear Alfven waves to compressible modes, in particular slow magnetosonic waves, cannot be ignored. This coupling pumps energy into the compressible modes, which then decay in shocks. The result has important implications for the decay of supersonic turbulence in the ISM of galaxies.

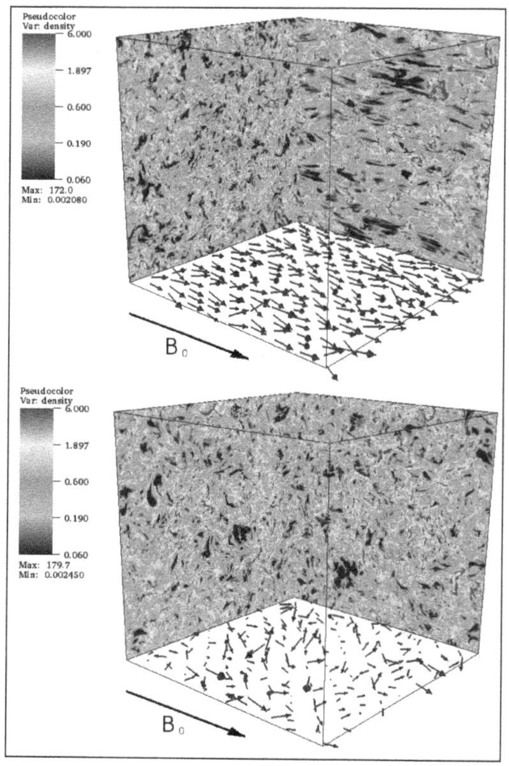

Structure of the density (grayscale) and magnetic field (arrows) in driven supersonic MHD turbulence for strong (top) and weak (bottom) fields.

Finally, more direct comparison between the simulations and observations is possible using properties such as the polarization angle of background star light. In many regions of the ISM, spinning dust grains become aligned with their long axis perpendicular to the magnetic field. When background stars are viewed through these aligned grains, their light is polarized, with the strength and direction of the polarization vector related to the column density of gas, and the magnetic field strength in the plane of the sky. Using numerical simulations of MHD turbulence, it is possible to compute theoretical maps of the polarization vectors along different viewing angles for background sources viewed through the simulation domain. Figure shows an example for two simulations, both using Mach 10 turbulence with strong ($\beta = 0.01$) and weak ($\beta = 1$) magnetic fields.

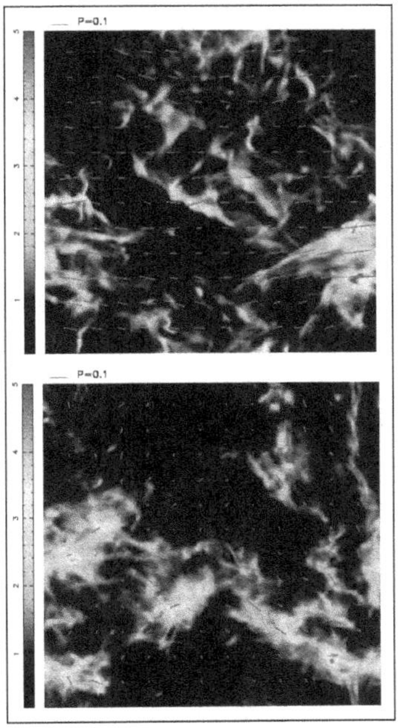

Scatter in polarization angle in supersonic turbulence with a strong field (top) and weak field (bottom). The grayscale shows the column density, and the line segments show the direction and amplitude of the polarization vector.

It is clear from inspection that in the case of strong fields, the scatter in polarization angle is small, while in the case of weak fields the scatter is large. In fact, this effect was predicted by Chandrasekhar & Fermi, who showed that the scatter in the polarization angle $\delta\varphi$ should be related to the plane-of-sky magnetic field strength B_p, gas density ρ, and line-of-sight velocity dispersion δv through,

$$B_p = 0.5\frac{(4\pi p)^{1/2}\,\delta v}{\delta\phi}$$

Equation above is now known as the "Chandrasekhar-Fermi" formula, and is now routinely used as a technique to measure magnetic field strengths in the ISM.

Kinetic MHD Effects in Clusters of Galaxies

Finally, we consider the effect of magnetic fields on the largest structures in the universe, clusters of galaxies. Radio observations of Faraday rotation in background sources indicate that the x-ray emitting plasma trapped in the gravitational potential of clusters is magnetized. Using the x-ray spectra to determine the temperature and density of the plasma shows that the mean free path of charged particles in the plasma is much smaller than the system size, but much larger than the gyroradius, that is the plasma is in the kinetic MHD regime.

The most important property of weakly collisional plasmas in the kinetic MHD regime, in comparison to highly collisional plasmas, is that the microscopic transport coefficients become anisotropic. For example, if the electron mean free path is much larger than the electron gyroradius, thermal conduction is primarily along magnetic field lines. Similarly, when the ion mean free path is much larger than the ion gyroradius, kinematic viscosity is primarily along magnetic field lines. The simplest description of the dynamics is therefore given by the equations of MHD supplemented by anisotropic thermal conduction and viscous transport terms.

Remarkably, the addition of anisotropic transport qualitatively changes the dynamics of the plasma. For example, with anisotropic thermal conduction, the convective stability criterion no longer depends on entropy, but only on the temperature gradient (if $dT/dz < 0$, the plasma is unstable to convection). Convective instability in this regime has been termed the magnetothermal instability (MTI). In fact, other instabilities have also been found in the kinetic MHD regime that might be important in clusters or in diffuse accretion flows.

Figure, taken directly from a nonlinear simulation, demonstrates the physics of the MTI. Consider a stratified atmosphere in a constant gravitational field. Arrange the vertical profiles of the pressure and density so that the atmosphere is hotter at the bottom than the top, and so that the entropy is constant or increasing upwards. In this case, the atmosphere are shown (which is hotter on the bottom than top), along with the direction of the heat flux Q induced along field lines which results in amplification of the perturbations should be stable to convection by the Schwarzschild criterion. Now consider a weak, horizontal magnetic field with anisotropic thermal conduction along field lines. Initially the field lines are parallel to the isotherms, so there is no heat flux in the equilibrium state. Now consider the evolution of vertical perturbations, as shown in the figure. The peaks of the perturbations are at a slightly lower pressure than their equilibrium position, so they expand and cool. The valleys are at a slightly higher pressure and so contract and heat up. These lead to a temperature gradient, and therefore a heat flux Q, along the field lines. The net result is to increase the entropy at the peaks (making them more buoyant), and to decrease the entropy at the valleys (making them sink). This increases the perturbation, tilts the field line more to the vertical, increases the temperature gradient along the field line and therefore increases the heat flux; and this process runs away as instability.

The nonlinear regime of the MTI has now been quite well studied using numerical simulations. With non-conducting boundaries at the top and bottom of the domain, the MTI saturates when the temperature profile becomes isothermal. If the top and bottom boundaries are held at fixed temperatures, then vigorous and sustained convection can be driven. How does the MTI relate to galaxy clusters? Recent work shows that it can play an important role in the temperature profiles of the x-ray emitting gas. When clusters form from gravitational collapse of large-scale structure, the initial temperature profile is centrally peaked. This profile is unstable to the MTI, and simulations of hydrostatic cores with weak magnetic fields show that the MTI causes significant redistribution of the temperature profile of the cluster, along with significant amplification of the magnetic field, in a Hubble time. More recently, the role that externally driven turbulence plays in the plasma dynamics, along with the MTI and other instabilities in the kinetic regime, has been an area of active inquiry.

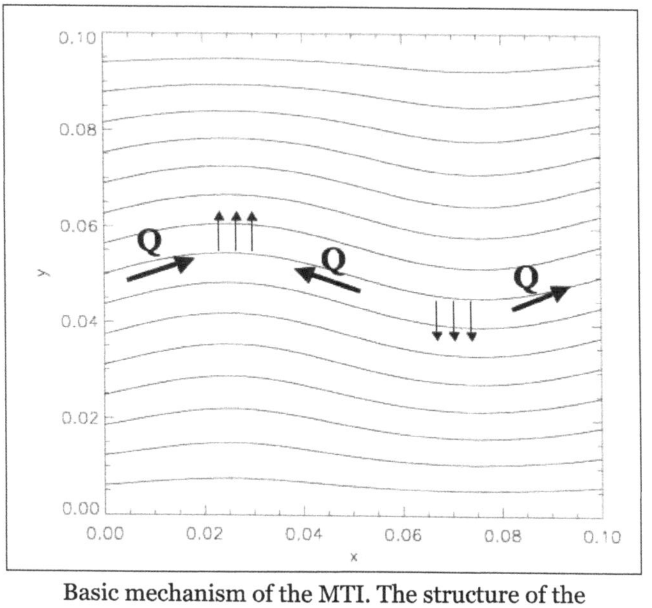

Basic mechanism of the MTI. The structure of the perturbed field lines in a stratified atmosphere.

Nuclear Astrophysics

Nuclear astrophysics is a field at the intersection of nuclear physics and astrophysics that seeks to understand how nuclear processes shape the cosmos. In essence we look for the connection between properties of atomic nuclei and the properties of planets, stars, and galaxies. Open questions include "How did the universe create the elements?", "How can extremely dense and hot astrophysical environments be used to learn about fundamental properties of matter?", and "How is the energy created that powers stars and stellar explosions?" One fascinating aspect of this field is its interdisciplinarity and diversity. Work in nuclear astrophysics includes astronomical observations using telescopes, gravitational wave detectors, and neutrino detectors; accelerator laboratory experiments using beams of stable nuclei, radioactive nuclei, neutrons, and gamma-rays; laboratory analysis of interstellar grains; large scale computer simulations of stellar explosions and nuclei; and theoretical work in nuclear physics and astrophysics.

Key Open Problems in the Field

Current key open questions include the question of the origin of the elements heavier than about germanium, in particular the elements from germanium to palladium that seem to have a much more complex origin than anticipated. Another open question is how we can use neutron stars as laboratories to learn about what happens with ordinary matter when it's compressed to very high densities. These ties into observations with X-ray observatories, radio telescopes, and the possible detection of gravitational waves from neutron star systems.

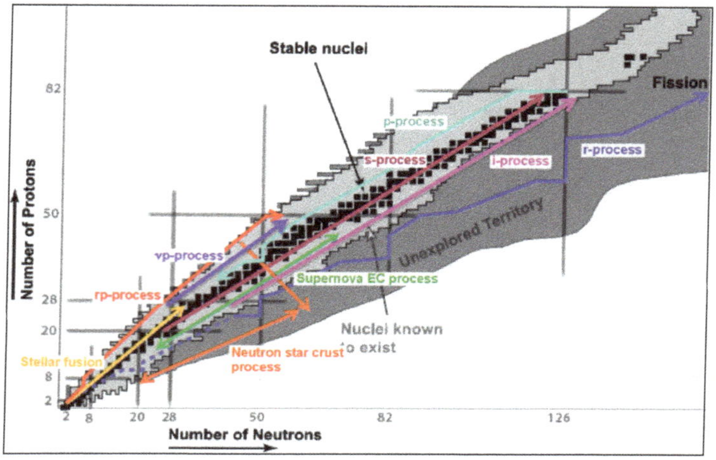

Nuclear processes in the Universe on the chart of nuclides.

The goals of nuclear astrophysics understand the energy generation in stars in all stages of stellar evolution and explaining the abundances of the elements and their isotopes as we observe them in nature. These aims are closely related since nuclear processes have been identified as the enormous energy source which stabilizes the stars and governs their evolution by transmuting nuclear species into other nuclear species thus simultaneously creating new elements.

Investigations during the last century have shown that we are connected to distant space and time not only by our imagination but also through a common cosmic heritage: the chemical elements that make up our bodies. These elements and their isotopes were created by nuclear fusion reactions in the hot interiors of remote and long-vanished stars over many billions of years. Their nuclear fuels finally spent, these giant stars met death in cataclysmic explosions, called supernovae (SN), scattering afar the atoms of heavy elements synthesized deep within their cores. Eventually this material, as well as material lost by smaller stars during the red-giant stages, collected into clouds of gas in interstellar space. These, in turn, slowly collapsed giving birth to new generations of stars, thus leading to a cycle of evolution that is still going on. In this scenario, the Sun and its complement of planets were formed some 5 billion years ago. The detailed understanding of the origin of the chemical elements combines astrophysics and nuclear physics, and forms what is called nuclear astrophysics.

Stars are powered by thermonuclear fusion reactions that convert protons into heavier chemical elements from He to Fe. The energy from these reactions amounts for the energy radiated from the stellar surface as well as the flux of stellar neutrinos.

Thermonuclear fusion reactions are at the heart of nuclear astrophysics. They sensitively influence the nucleosynthesis of the elements in the earliest stages of the universe and in all the objects formed thereafter, and control the associated energy generation, neutrino luminosity, and evolution of stars. A good knowledge of the rates of these reactions is thus essential for understanding the broad picture of stellar evolution. Nuclear astrophysics has entered a period characterized by unprecedented demands for new information and for improved accuracy. This challenge is being driven by the quality and breadth of current observational data, a direct consequence of impressive advances in observational technology. The extraordinary results provided by the satellite telescopes INTEGRAL and HUBBLE or the precise maps of the cosmic microwave background background provided by PLANCK are just some examples.

Observations of elemental and isotopic abundances are very useful probes of stellar structure and evolution. In some cases, these abundances depend sensitively on the state of the stellar interior and, provided that the relevant nuclear information exists, they allow us to construct detailed models of stars. At other times, the observed abundance falls entirely outside the predictions of stellar models and requires new ideas before being incorporated into the general framework of nucleosynthesis. However, measurements of astrophysically-interesting reactions are complicated by the fact that reaction rates at stellar temperatures are almost vanishingly small.

Moreover, low energy studies of thermonuclear reactions in a laboratory at the earth's surface are hampered by background effects of cosmic rays in the detectors, for that the best solution is to install the dedicated accelerator facility in a deep underground laboratory, where the rock shield against cosmic rays.

Computational Astrophysics

Computational astrophysics is the use of numerical methods to solve research problems in astrophysics on a computer.

Numerical methods are used whenever the mathematical model describing an astrophysical system is too complex to solve analytically (with pencil and paper). Today, it is difficult to find examples of research problems that do not use computation.

Computational vs. Analytic Methods

Solutions generated by numerical methods are generally only approximations to the exact solution of the underlying equations. However, much more complex systems of equations can be solved numerically than can be solved analytically. Thus, approximate solutions to the exact equations found by numerical methods often provide far more insight than exact solutions to approximate equations that can be solved analytically. For example, time-dependent numerical solutions of fluid flow in three dimensions can exhibit behavior which is not expected from one-dimensional analytic solutions to the steady-state (time independent) equations.

The increase in computing power in the last few decades has meant that an increasingly larger share of problems in astrophysics can be solved on a desktop computer. However, the most computationally intensive problems in astrophysics are still limited by the memory and floating-point speed of the largest high-performance computer (HPC) systems available. The use of HPC is necessary to maximize the spatial, temporal, or frequency resolution of solutions, to include more physics, or when large parameter surveys are required to understand the statistics.

Examples:

Traditionally, the most important applications of computation in astrophysics have been in the areas of:

- Stellar structure and evolution: The internal structure and evolution of stars with different masses and chemical composition was largely mapped out in the 1960s using numerical methods to solve the equations of stellar structure. Today, the frontiers of research include calculating multidimensional stellar models of rapidly rotating stars, modeling the effects of hydrodynamical processes such as convection from first principles, and understanding how stars generate magnetic fields through dynamo processes.
- Radiation transfer and stellar atmospheres: Without oversimplifying assumptions, computational methods are required to calculate the propagation of light through the outer layers of a star, including its interaction with matter through absorption, emission, and scattering of photons. The calculation of cross sections for the interaction of light with matter for astrophysically relevant ions is itself a challenging computational problem. The construction of such stellar atmosphere models share many challenges with radiation transfer problems in other systems such as planets, accretion disks, and the interstellar medium. Modern calculations improve the frequency resolution, include a better treatment of opacities, and can treat non-hydrostatic atmospheres.
- Astrophysical fluid dynamics: The dynamics of most of the visible matter in the universe can be treated as a compressible fluid. Time-dependent and multidimensional solutions to the fluid equations, including the effects of gravitational, magnetic, and radiation fields, require numerical methods. A vast range of problems are addressed in this way, from convection and dynamo action in

stellar and planetary interiors, to the formation of galaxies and the large scale structure of the universe.

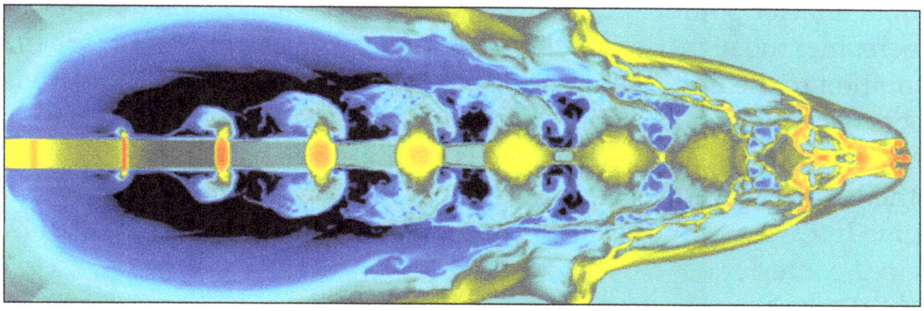

Density in a 2D simulations of the propagation of a pulsed, supersonic, protostellar jet.

- Planetary, stellar, and galactic dynamics: It is well known that Newton's Laws of Motion for some number of particles N interacting through their mutual gravitational attraction do not in general have an analytic solution for $N > 2$. Thus, to compute the orbits of planets in the solar system, or of stars in the Galaxy, numerical methods are required. The most challenging problems today include accurate integration of the orbits of the planets over the age of the solar system, studying the dynamics of globular clusters including the effect of stellar evolution and the formation of binaries, studying galaxy mergers and interaction, and computing structure formation in the universe through the gravitational clustering of collisionless dark matter.

Numerical Methods

The diverse set of mathematical models encountered in astrophysics means that a very wide range of numerical methods are necessary. These range from basic methods for linear algebra, nonlinear root finding, and ordinary differential equations (ODEs), to more complex methods for coupled partial differential equations (PDEs) in multi-dimensions, topics which span the entire content of reference books such as Numerical Recipes.

However, there are several numerical methods used in astrophysics that deserve special mention, either because they have such wide use in astrophysics, or because astrophysicists have made significant contributions to their development. These methods are considered in the following subsections.

Stellar Structure Codes

The equations of stellar structure are a set of ODEs which define a classic two-point boundary value problem. Analytic solutions to an approximate system of equations exist in special cases (polytropes) but these are of limited application to real stars. Numerical solutions to the full system of equations were first computed using shooting methods, in which boundary conditions are guessed at the center and surface of the star, the equations are integrated outwards and inwards, and matching conditions are used at some interior point to select the full solution. Shooting methods are laborious and cumbersome; today modern stellar structure codes use relaxation schemes which find the solution to the finite-difference form of the stellar structure equations by finding the roots of coupled, non-linear equations at each mesh point. A good example of a public code that uses a relaxation scheme is the EZ-code, based on Eggleton's variable mesh method. The evolution of stars are then computed by computing stellar models at discrete time intervals, with the chemical composition of the star modified by nuclear reactions in the interior.

Radiative Transfer Codes

Calculating the emergent intensity from an astrophysical system requires solving a multidimensional integral-differential equation, along with level population equations to account for the interaction of the radiation with matter. In general the solution is a function of two angles, frequency, and time. Even in static, plane-parallel atmospheres, the problem is two-dimensional (one angle and frequency). However, the most challenging aspect of the problem is that scattering couples the solutions at different angles and frequencies. As in the stellar structure problem, relaxation schemes are used to solve the finite difference form of the transfer equations, although specialized iteration techniques are necessary to accelerate convergence. Monte-Carlo methods, which adopt statistical techniques to approximate the solution by following the propagation of many photon packets, are becoming increasingly important. The problem of line-transfer in a moving atmosphere (stellar wind) is especially challenging, due to non-local coupling introduced by Doppler shifts in the spectrum.

N-body Codes

There are two tasks in an N-body code: integrating the equations of motion (pushing particles), and computing the gravitational acceleration of each particle. The former requires methods for integrating ODEs. Modern codes are based on a combination of high-order difference approximations (e.g. Hermite integrators), and symplectic methods (which have the important property of generating solutions that obey Liouville's Theorem, i.e. that preserve the volume of the solution in phase space). Symplectic methods are especially important for long term time integration, because they control the accumulation of truncation error in the solution.

Calculation of the gravitational acceleration is challenging because the computational cost scales as N(N-1), where N is the number of particles. For small N, direct summation can be used. For moderate N (currently $N\sim10^{5-6}$), special purpose hardware (e.g. GRAPE boards) can be used to accelerate the evaluation of $1/r^2$ necessary to compute the acceleration by direct summation. Finally, for large N (currently $N\geq10^{9-10}$), tree-methods are used to approximate the force from distant particles.

Codes for Astrophysical Fluid Dynamics

Solving the equations of compressible gas dynamics is a classic problem in numerical analysis which has application to many fields besides astrophysics. Thus, a large number of methods have been developed, with many important contributions being made by astrophysicists. For solving the equations of compressible fluid dynamics, the most popular methods include:

- Finite-difference techniques (which require hyper-viscosity to smooth discontinuities).
- Finite-volume methods (which often use a riemann solver to compute upwind fluxes).
- Operator split methods (which combine elements of both finite-differencing and finite-volume methods for different terms in the equations).
- Central methods (which often use simple expressions for the fluxes, combined with high-order spatial interpolation).
- Particle methods such as smooth particle hydrodynamics (SPH, which integrates the motion of discrete particles to follow the flow).

SPH is an example of a method developed largely to solve astrophysics problems, although many of the developments in other methods (e.g. the extension of finite-difference and finite-volume methods to magnetohydrodynamics (MHD) to include the effects of magnetic fields on the dynamics of the fluid) have also been motivated by astrophysics.

Relation to other Fields

Computational astrophysics is necessarily inter-disciplinary, involving aspects of not only astrophysics, but also numerical analysis and computer science.

Numerical Analysis

Numerical Analysis is a rigorous branch of mathematics concerned with the approximation of functions and integrals, and the approximation of solutions to algebraic, differential, and integral equations. It provides tools to analyze errors that arise from the approximations themselves (truncation error), and from the use of finite-precision arithmetic on a computer (round-off error). Convergence, consistency, and stability of numerical algorithms are all essential for their use in practical applications. Thus, the development of new numerical algorithms to solve problems in astrophysics is deeply rooted in the tools of numerical analysis.

Computer Science

Computational science and computer science differ. The former is the use of numerical methods to solve scientific problems. The latter is the study of computers and computation. Thus, computer science tends to be more orientated towards the theory of computers and computation, while scientific computation is concerned with the practical aspects of solving science problems. Still, there is a large degree of overlap between the fields, with many computer scientists engaged in developing tools for scientific computation, and many scientists working on software problems of interest to computer scientists. Examples of the overlap include the development of standards for parallel processing such as the Message Passing Interface (MPI) and OpenMP, and development of parallel I/O filesystems such as Lustre.

References

- Celestial-mechanics-physics, science: britannica.com, Retrieved 29 April, 2019
- Keplers-laws-of-planetary-motion, science: britannica.com, Retrieved 06 May, 2019
- Astrophysical-plasma: plasma-universe.com, Retrieved 11 March, 2019
- Hydromagnetic-dynamo-theory: scholarpedia.org, Retrieved 07 June, 2019
- Computational-astrophysics: scholarpedia.org, Retrieved 11 March, 2019

2

Stellar Astrophysics

Stellar astrophysics is one of the sub-disciplines of astrophysics which studies the structure and properties of stars. It includes star formation, stellar luminosities, stellar parallax, stellar positions, stellar magnetic fields, nova spectra, etc. This chapter explores these aspects of stellar astrophysics to provide an extensive understanding of the subject.

Stellar astrophysics is the branch of astronomy which studies the evolution of stellar structure and their oscillations. This study of evolution of stars helps the study of evolution of the universe because the energy output of the formation of stars are added evolution of galaxies, as they are closely coupled with planet formation as well.

Another use of star as a probe is detecting the dark matter gravitational effects. This field of study is one of the major disciplines of experimental astronomy. This helps to study about the big bang theory and early time of universe.

In the beginning of big bang, hydrogen and helium were resulted all over the galaxy but now different planets with different elements exist. Earth contains life forms since heavier elements such as oxygen, nitrogen carbon etc. are formed by fusion reactions in the interior of stars.

Stellar astrophysics discusses on these reactions ng the difference in first and later evolved stars, supernovas and production of heavier elements that even supports life forms. This study includes the conditions which are essential to the formation of a star and the properties of the stars as their brightness and energy output. The interactions between galaxies are also one of the outputs of the stellar astrophysics.

This study involves the stellar model as a part of it. This helps to study the internal structure and mechanisms which results the determination of color and luminosity of that star. Accurate model will help not only the study of certain star, but also the study of brown dwarfs, supernovae and more.

Stars

Star is any massive self-luminous celestial body of gas that shines by radiation derived from its internal energy sources. Of the tens of billions of trillions of stars composing the observable universe, only a very small percentage is visible to the naked eye. Many stars occur in pairs, multiple systems, or star clusters. The members of such stellar groups are physically related through common origin and are bound by mutual gravitational attraction. Somewhat related to star clusters are stellar associations, which consist of loose groups of physically similar stars that have insufficient mass as a group to remain together as an organization.

Open cluster NGC 290. Stars in the open cluster NGC 290, as seen by the Hubble Space Telescope.

The Sun as a Point of Comparison

Variations in Stellar Size

With regard to mass, size, and intrinsic brightness, the Sun is a typical star. Its approximate mass is 2×10^{30} kg (about 330,000 Earth masses), its approximate radius 700,000 km (430,000 miles), and its approximate luminosity 4×10^{33} ergs per second (or equivalently 4×10^{23} kilowatts of power). Other stars often have their respective quantities measured in terms of those of the Sun.

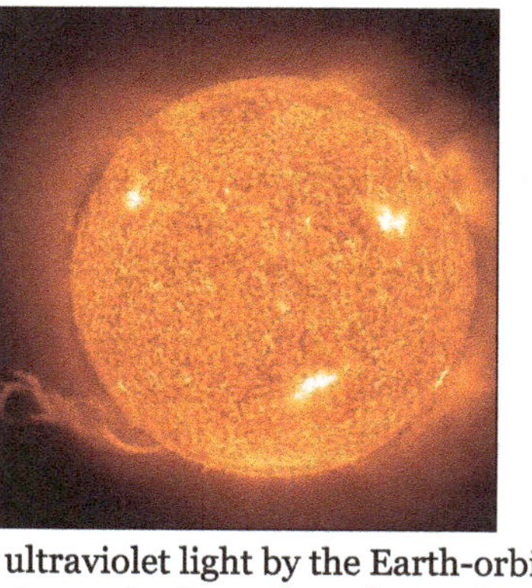

The Sun as imaged in extreme ultraviolet light by the Earth-orbiting Solar and Heliospheric Observatory (SOHO) satellite. A massive loop-shaped eruptive prominence is visible at the lower left. Nearly white areas are the hottest; deeper reds indicate cooler temperatures.

Many stars vary in the amount of light they radiate. Stars such as Altair, Alpha Centauri A and B, and Procyon A are called dwarf stars; their dimensions are roughly comparable to those of the Sun. Sirius A and Vega, though much brighter, also are dwarf stars; their higher temperatures yield a larger rate of emission per unit area. Aldebaran A,

Arcturus, and Capella A are examples of giant stars, whose dimensions are much larger than those of the Sun. Observations with an interferometer (an instrument that measures the angle subtended by the diameter of a star at the observer's position), combined with parallax measurements (which yield a star's distance, give sizes of 12 and 22 solar radii for Arcturus and Aldebaran A. Betelgeuse and Antares A are examples of supergiant stars. The latter has a radius some 300 times that of the Sun, whereas the variable star Betelgeuse oscillates between roughly 300 and 600 solar radii. Several of the stellar class of white dwarf stars, which have low luminosities and high densities, also are among the brightest stars. Sirius B is a prime example, having a radius one-thousandth that of the Sun, which is comparable to the size of Earth. Also among the brightest stars are Rigel A, a young supergiant in the constellation Orion, and Canopus, a bright beacon in the Southern Hemisphere often used for spacecraft navigation.

Stellar Activity and Mass Loss

The Sun's activity is apparently not unique. It has been found that stars of many types are active and have stellar winds analogous to the solar wind. The importance and ubiquity of strong stellar winds became apparent only through advances in spaceborne ultraviolet and X-ray astronomy as well as in radio and infrared surface-based astronomy.

X-ray observations that were made during the early 1980s yielded some rather unexpected findings. They revealed that nearly all types of stars are surrounded by coronas having temperatures of one million kelvins (K) or more. Furthermore, all stars seemingly display active regions, including spots, flares, and prominences much like those of the Sun. Some stars exhibit starspots so large that an entire face of the star is relatively dark, while others display flare activity thousands of times more intense than that on the Sun.

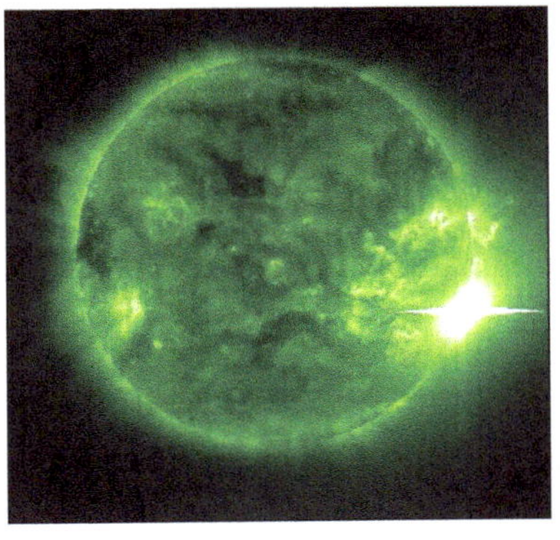

Solar flare: One of the strongest solar flares ever detected, in an extreme ultraviolet (false-colour) image of the Sun taken by the Solar and Heliospheric Observatory (SOHO) satellite, November 4, 2003. Such powerful flares, called X-class flares, release intense radiation that can temporarily cause blackouts in radio communications all over Earth.

The highly luminous hot, blue stars have by far the strongest stellar winds. Observations of their ultraviolet spectra with telescopes on sounding rockets and spacecraft have shown that their wind speeds often reach 3,000 km (roughly 2,000 miles) per second, while losing mass at rates up to a billion times that of the solar wind. The corresponding mass-loss rates approach and sometimes exceed one hundred-thousandth of a solar mass per year, which means that one entire solar mass (perhaps a tenth of the total mass of the star) is carried away into space in a relatively short span of 100,000 years. Accordingly, the most luminous stars are thought to lose substantial fractions of their mass during their lifetimes, which are calculated to be only a few million years.

Ultraviolet observations have proved that to produce such great winds the pressure of hot gases in a corona, which drives the solar wind, is not enough. Instead, the winds of the hot stars must be driven directly by the pressure of the energetic ultraviolet radiation emitted by these stars. Aside from the simple realization that copious quantities of ultraviolet radiation flow from such hot stars, the details of the process are not well understood. Whatever is going on, it is surely complex, for the ultraviolet spectra of the stars tend to vary with time, implying that the wind is not steady. In an effort to understand better the variations in the rate of flow, theorists are investigating possible kinds of instabilities that might be peculiar to luminous hot stars.

Observations made with radio and infrared telescopes as well as with optical instruments prove that luminous cool stars also have winds whose total mass-flow rates are comparable to those of the luminous hot stars, though their velocities are much lower—about 30 km (20 miles) per second. Because luminous red stars are inherently cool objects (having a surface temperature of about 3,000 K, or half that of the Sun), they emit very little detectable ultraviolet or X-ray radiation; thus, the mechanism driving the winds must differ from that in luminous hot stars. Winds from luminous cool stars, unlike those from hot stars, are rich in dust grains and molecules. Since nearly all stars more massive than the Sun eventually evolve into such cool stars, their winds, pouring into space from vast numbers of stars, provide a major source of new gas and dust in interstellar space, thereby furnishing a vital link in the cycle of star formation and galactic evolution. As in the case of the hot stars, the specific mechanism that drives the winds of the cool stars is not understood; at this time, investigators can only surmise that gas turbulence, magnetic fields, or both in the atmospheres of these stars are somehow responsible.

Strong winds also are found to be associated with objects called protostars, which are huge gas balls that have not yet become full-fledged stars in which energy is provided by nuclear reactions. Radio and infrared observations of deuterium (heavy hydrogen) and carbon monoxide (CO) molecules in the Orion Nebula have revealed clouds of gas expanding outward at velocities approaching 100 km (60 miles) per second. Furthermore, high-resolution, very-long-baseline interferometry observations have disclosed expanding knots of natural maser (coherent microwave) emission of water vapour near the star-forming regions in Orion, thus linking the strong winds to the protostars themselves. The specific causes of these winds remain unknown, but if they generally accompany star formation, astronomers will have to consider the implications for the early solar system. After all, the Sun was presumably once a protostar too.

Distances to the Stars

Determining Stellar Distances

Distances to stars were first determined by the technique of trigonometric parallax, a method still used for nearby stars. When the position of a nearby star is measured from two points on opposite sides of Earth's orbit (i.e., six months apart), a small angular (artificial) displacement is observed relative to a background of very remote (essentially fixed) stars. Using the radius of Earth's orbit as the baseline, the distance of the star can be found from the parallactic angle, p. If p = 1″ (one second of arc), the distance of the star is 206,265 times Earth's distance from the Sun—namely, 3.26 light-years. This unit of distance is termed the parsec, defined as the distance of an object whose parallax equals one arc second. Therefore, one parsec equals 3.26 light-years. Since parallax is inversely proportional to distance, a star at 10 parsecs would have a parallax of 0.1″. The nearest star to Earth, Proxima Centauri (a member of the triple system of Alpha Centauri), has a parallax of 0.76813″, meaning that its distance is 1/0.76813, or 1.302, parsecs, which equals 4.24 light-years. The parallax of Barnard's star, the next closest after the Alpha Centauri system, is 0.54831″, so that its distance is nearly 6 light-years. Errors of such parallaxes are now typically 0.001″. Thus, measurements of trigonometric parallaxes are useful for only the nearby stars within a few thousand light-years. In fact, of the approximately 100 billion stars in the Milky Way Galaxy (also simply called the Galaxy), the Hipparcos satellite has measured only about 100,000 to an accuracy of 0.001″. For more distant stars, indirect methods are used; most of them depend on comparing the intrinsic brightness of a star (found, for example, from its spectrum or other observable property) with its apparent brightness.

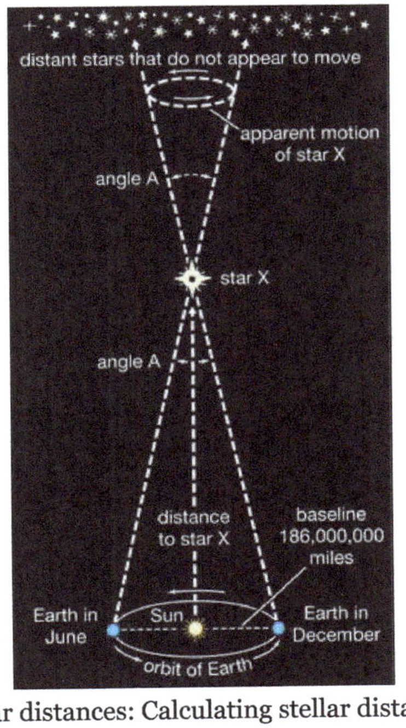

Stellar distances: Calculating stellar distances.

Nearest Stars

Only three stars, Alpha Centauri, Procyon, and Sirius, are both among the 20 nearest and among the 20 brightest stars. Ironically, most of the relatively nearby stars are dimmer than the Sun and are invisible without the aid of a telescope. By contrast, some of the well-known bright stars outlining the constellations have parallaxes as small as the limiting value of 0.001″ and are therefore well beyond several hundred light-years' distance from the Sun. The most luminous stars can be seen at great distances, whereas the intrinsically faint stars can be observed only if they are relatively close to Earth.

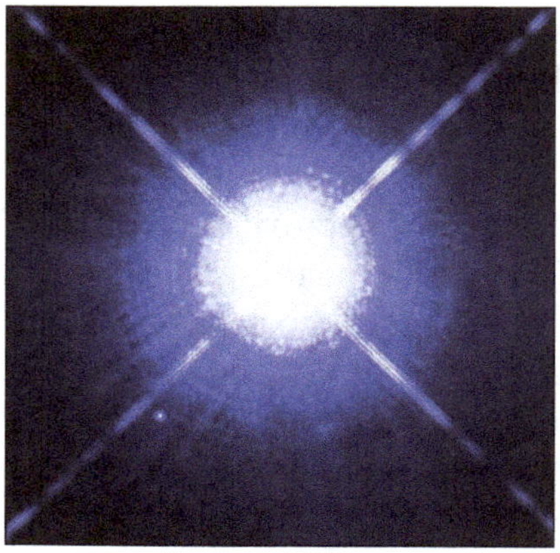

Sirius: Sirius A and B (lower left) photographed by the Hubble Space Telescope.

Although the lists of the brightest and the nearest stars pertain to only a very small number of stars, they nonetheless serve to illustrate some important points. The stars listed fall roughly into three categories: (1) giant stars and supergiant stars having sizes of tens or even hundreds of solar radii and extremely low average densities—in fact, several orders of magnitude less than that of water (one gram per cubic centimetre); (2) dwarf stars having sizes ranging from 0.1 to 5 solar radii and masses from 0.1 to about 10 solar masses; and (3) white dwarf stars having masses comparable to that of the Sun but dimensions appropriate to planets, meaning that their average densities are hundreds of thousands of times greater than that of water.

These rough groupings of stars correspond to stages in their life histories. The second category is identified with what is called the main sequence and includes stars that emit energy mainly by converting hydrogen into helium in their cores. The first category comprises stars that have exhausted the hydrogen in their cores and are burning hydrogen within a shell surrounding the core. The white dwarfs represent the final stage in the life of a typical star, when most available sources of energy have been exhausted and the star has become relatively dim.

The large number of binary stars and even multiple systems is notable. These star systems exhibit scales comparable in size to that of the solar system. Some, and perhaps many, of the nearby single stars have invisible (or very dim) companions detectable by their gravitational effects on the primary star; this orbital motion of the unseen member causes the visible star to "wobble" in its motion through space. Some of the invisible companions have been found to have masses on the order of 0.001 solar mass or less, which is in the range of planetary rather than stellar dimensions. Current observations suggest that they are genuine planets, though some are merely extremely dim stars (sometimes called brown dwarfs). Nonetheless, a reasonable inference that can be drawn from these data is that double stars and planetary systems are formed by similar evolutionary processes.

Stellar Magnitudes

Measuring Starlight Intensity

Stellar brightnesses are usually expressed by means of their magnitudes, a usage inherited from classical times. A star of the first magnitude is about 2.5 times as bright as one of the second magnitude, which in turn is some 2.5 times as bright as one of the third magnitude, and so on. A star of the first magnitude is therefore 2.5^5 or 100 times as bright as one of the sixth magnitude. The magnitude of Sirius, which appears to an observer on Earth as the brightest star in the sky (save the Sun), is −1.4. Canopus, the second brightest, has a magnitude of −0.7, while the faintest star normally seen without the aid of a telescope is of the sixth magnitude. Stars as faint as the 30th magnitude have been measured with modern telescopes, meaning that these instruments can detect stars about four billion times fainter than can the human eye alone.

The scale of magnitudes comprises a geometric progression of brightness. Magnitudes can be converted to light ratios by letting l_n and l_m be the brightnesses of stars of magnitudes n and m; the logarithm of the ratio of the two brightnesses then equals 0.4 times the difference between them—i.e., $\log(l_m/l_n) = 0.4(n - m)$. Magnitudes are actually defined in terms of observed brightness, a quantity that depends on the light-detecting device employed. Visual magnitudes were originally measured with the eye, which is most sensitive to yellow-green light, while photographic magnitudes were obtained from images on old photographic plates, which were most sensitive to blue light. Today, magnitudes are measured electronically, using detectors such as CCDs equipped with yellow-green or blue filters to create conditions that roughly correspond to those under which the original visual and photographic magnitudes were measured. Yellow-green magnitudes are still often designated V magnitudes, but blue magnitudes are now designated B. The scheme has been extended to other magnitudes, such as ultraviolet (U), red (R), and near-infrared (I). Other systems vary the details of this scheme. All magnitude systems must have a reference, or zero, point. In practice, this is fixed arbitrarily by agreed-upon magnitudes measured for a variety of standard stars.

The actually measured brightness's of stars give apparent magnitudes. These cannot be converted to intrinsic brightness's until the distances of the objects concerned are known. The absolute magnitude of a star is defined as the magnitude it would have if it were viewed at a standard distance of 10 parsecs (32.6 light-years). Since the apparent visual magnitude of the Sun is −26.75, its absolute magnitude corresponds to a diminution in brightness by a factor of $(2,062,650)^2$ and is, using logarithms, $-26.75 + 2.5 \times \log(2,062,650)^2$, or $-26.75 + 31.57 = 4.82$. This is the magnitude that the Sun would have if it were at a distance of 10 parsecs—an object still visible to the naked eye, though not a very conspicuous one and certainly not the brightest in the sky. Very luminous stars, such as Deneb, Rigel, and Betelgeuse, have absolute magnitudes of −7 to −9, while one of the faintest known stars, the companion to the star with the catalog name BD + 4°4048, has an absolute visual magnitude of +19, which is about a million times fainter than the Sun. Many astronomers suspect that large numbers of such faint stars exist, but most of these objects have so far eluded detection.

Stellar Colours

Stars differ in colour. Most of the stars in the constellation Orion visible to the naked eye are blue-white, most notably Rigel (Beta Orionis), but Betelgeuse (Alpha Orionis) is a deep red. In the telescope, Albireo (Beta Cygni) is seen as two stars, one blue and the other orange. One quantitative means of measuring stellar colours involves a comparison of the yellow (visual) magnitude of the star with its magnitude measured through a blue filter. Hot, blue stars appear brighter through the blue filter, while the opposite is true for cooler, red stars. In all magnitude scales, one magnitude step corresponds to a brightness ratio of 2.512. The zero point is chosen so that white stars with surface temperatures of about 10,000 K have the same visual and blue magnitudes. The conventional colour index is defined as the blue magnitude, B, minus the visual magnitude, V; the colour index, B − V, of the Sun is thus $+5.47 - 4.82 = 0.65$.

Orion: The constellation Orion is one of the easiest to recognize because of a group of three stars. The three stars form a straight line that is often called Orion's Belt. The Orion Nebula can be seen as a pink fuzzy light below the line of three stars. The red star Betelgeuse is in the upper left, and the bright star Rigel is in the lower right.

Magnitude Systems

Problems arise when only one colour index is observed. If, for instance, a star is found to have, say, a B – V colour index of 1.0 (i.e., a reddish colour), it is impossible without further information to decide whether the star is red because it is cool or whether it is really a hot star whose colour has been reddened by the passage of light through interstellar dust. Astronomers have overcome these difficulties by measuring the magnitudes of the same stars through three or more filters, often U (ultraviolet), B, and V.

Observations of stellar infrared light also have assumed considerable importance. In addition, photometric observations of individual stars from spacecraft and rockets have made possible the measurement of stellar colours over a large range of wavelengths. These data are important for hot stars and for assessing the effects of interstellar attenuation.

Bolometric Magnitudes

The measured total of all radiation at all wavelengths from a star is called a bolometric magnitude. The corrections required reducing visual magnitudes to bolometric magnitudes are large for very cool stars and for very hot ones, but they are relatively small for stars such as the Sun. A determination of the true total luminosity of a star affords a measure of its actual energy output. When the energy radiated by a star is observed from Earth's surface, only that portion to which the energy detector is sensitive and that can be transmitted through the atmosphere is recorded. Most of the energy of stars like the Sun is emitted in spectral regions that can be observed from Earth's surface. On the other hand, a cool dwarf star with a surface temperature of 3,000 K has an energy maximum on a wavelength scale at 10000 angstroms (Å) in the far-infrared, and most of its energy cannot therefore be measured as visible light. (One angstrom equals 10^{-10} metre, or 0.1 nanometre.) Bright, cool stars can be observed at infrared wavelengths, however, with special instruments that measure the amount of heat radiated by the star. Corrections for the heavy absorption of the infrared waves by water and other molecules in Earth's air must be made unless the measurements are made from above the atmosphere.

The hotter stars pose more difficult problems, since Earth's atmosphere extinguishes all radiation at wavelengths shorter than 2900 Å. A star whose surface temperature is 20,000 K or higher radiates most of its energy in the inaccessible ultraviolet part of the electromagnetic spectrum. Measurements made with detectors flown in rockets or spacecraft extend the observable wavelength region down to 1000 Å or lower, though most radiation of distant stars is extinguished below 912 Å—a region in which absorption by neutral hydrogen atoms in intervening space becomes effective.

To compare the true luminosities of two stars, the appropriate bolometric corrections must first be added to each of their absolute magnitudes. The ratio of the luminosities can then be calculated.

Stellar Spectra

A star's spectrum contains information about its temperature, chemical composition, and intrinsic luminosity. Spectrograms secured with a slit spectrograph consist of a sequence of images of the slit in the light of the star at successive wavelengths. Adequate spectral resolution (or dispersion) might show the star to be a member of a close binary system, in rapid rotation, or to have an extended atmosphere. Quantitative determination of its chemical composition then becomes possible. Inspection of a high-resolution spectrum of the star may reveal evidence of a strong magnetic field.

Line Spectrum

Spectral lines are produced by transitions of electrons within atoms or ions. As the electrons move closer to or farther from the nucleus of an atom (or of an ion), energy in the form of light (or other radiation) is emitted or absorbed. The yellow D lines of sodium or the H and K lines of ionized calcium (seen as dark absorption lines) are produced by discrete quantum jumps from the lowest energy levels (ground states) of these atoms. The visible hydrogen lines however, are produced by electron transitions within atoms in the second energy level (or first excited state), which lies well above the ground level in energy. Only at high temperatures are sufficient numbers of atoms maintained in this state by collisions, radiations, and so forth to permit an appreciable number of absorptions to occur. At the low surface temperatures of a red dwarf star, few electrons populate the second level of hydrogen, and thus the hydrogen lines are dim. By contrast, at very high temperatures—for instance, that of the surface of a blue giant star—the hydrogen atoms are nearly all ionized and therefore cannot absorb or emit any line radiation. Consequently, only faint dark hydrogen lines are observed. The characteristic features of ionized metals such as iron are often weak in such hotter stars because the appropriate electron transitions involve higher energy levels that tend to be more sparsely populated than the lower levels. Another factor is that the general "fogginess," or opacity, of the atmospheres of these hotter stars is greatly increased, resulting in fewer atoms in the visible stellar layers capable of producing the observed lines.

The continuous (as distinct from the line) spectrum of the Sun is produced primarily by the photodissociation of negatively charged hydrogen ions (H^-)—i.e., atoms of hydrogen to which an extra electron is loosely attached. In the Sun's atmosphere, when H^- is subsequently destroyed by photodissociation, it can absorb energy at any of a whole range of wavelengths and thus produce a continuous range of absorption of radiation. The main source of light absorption in the hotter stars is the photoionization of hydrogen atoms, both from ground level and from higher levels.

Spectral Analysis

The physical processes behind the formation of stellar spectra are well enough understood to permit determinations of temperatures, densities, and chemical compositions of stellar atmospheres. The star studied most extensively is, of course, the Sun.

The general characteristics of the spectra of stars depend more on temperature variations among the stars than on their chemical differences. Spectral features also depend on the density of the absorbing atmospheric matter, and density in turn is related to a star's surface gravity. Dwarf stars, with great surface gravities, tend to have high atmospheric densities; giants and supergiants, with low surface gravities, have relatively low densities. Hydrogen absorption lines provide a case in point. Normally, an undisturbed atom radiates a very narrow line. If its energy levels are perturbed by charged particles passing nearby, it radiates at a wavelength near its characteristic wavelength. In a hot gas, the range of disturbance of the hydrogen lines is very high, so that the spectral line radiated by the whole mass of gas is spread out considerably; the amount of blurring depends on the density of the gas in a known fashion. Dwarf stars such as Sirius show broad hydrogen features with extensive "wings" where the line fades slowly out into the background, while supergiant stars, with less-dense atmospheres, display relatively narrow hydrogen lines.

Stellar Temperatures

Temperatures of stars can be defined in a number of ways. From the character of the spectrum and the various degrees of ionization and excitation found from its analysis, an ionization or excitation temperature can be determined.

A comparison of the V and B magnitudes yields a B – V colour index, which is related to the colour temperature of the star. The colour temperature is therefore a measure of the relative amounts of radiation in two more or less broad wavelength regions, while the ionization and excitation temperatures pertain to the temperatures of strata wherein spectral lines are formed.

Provided that the angular size of a star can be measured and that the total energy flux received at Earth (corrected for atmospheric extinction) is known, the so-called brightness temperature can be found.

The effective temperature, T_{eff}, of a star is defined in terms of its total energy output and radius. Thus, since σT^4_{eff} is the rate of radiation per unit area for a perfectly radiating sphere and if L is the total radiation (i.e., luminosity) of a star considered to be a sphere of radius R, such a sphere (called a blackbody) would emit a total amount of energy equal to its surface area, $4\pi R^2$, multiplied by its energy per unit area. In symbols, $L = 4\pi R^2 \sigma T^4_{eff}$. This relation defines the star's equivalent blackbody, or effective, temperature.

Since the total energy radiated by a star cannot be directly observed (except in the case of the Sun), the effective temperature is a derived quantity rather than an observed one. Yet, theoretically, it is the fundamental temperature. If the bolometric corrections are known, the effective temperature can be found for any star whose absolute visual magnitude and radius are known. Effective temperatures are closely related to spectral type and range from about 40,000 K for hot O-type stars, through 5,800 K for stars like the Sun, to about 300 K for brown dwarfs.

Stellar Masses

Masses of stars can be found directly only from binary systems and only if the scale of the orbits of the stars around each other is known. Binary stars are divided into three categories, depending on the mode of observation employed: visual binaries, spectroscopic binaries, and eclipsing binaries.

Visual Binaries

Visual binaries can be seen as double stars with the telescope. True doubles, as distinguished from apparent doubles caused by line-of-sight effects, move through space together and display a common space motion. Sometimes a common orbital motion can be measured as well. Provided that the distance to the binary is known, such systems permit a determination of stellar masses, m_1 and m_2, of the two members. The angular radius, a″, of the orbit (more accurately, its semimajor axis) can be measured directly, and, with the distance known, the true dimensions of the semimajor axis, a, can be found. If a is expressed in astronomical units, which is given by a (measured in seconds of arc) multiplied by the distance in parsecs, and the period, P, also measured directly, is expressed in years, then the sum of the masses of the two orbiting stars can be found from an application of Kepler's third law. (An astronomical unit is the average distance from Earth to the Sun, 149,597,870.7 km [92,955,807.3 miles].) In symbols, $(m_1 + m_2) = a^3/P^2$ in units of the Sun's mass. For example, for the binary system 70 Ophiuchi, P is 87.8 years, and the distance is 5.0 parsecs; thus, a is 22.8 astronomical units, and $m_1 + m_2 = 1.56$ solar masses. From a measurement of the motions of the two members relative to the background stars, the orbit of each star has been determined with respect to their common centre of gravity. The mass ratio, $m_2/(m_1 + m_2)$, is 0.42; the individual masses for m_1 and m_2, respectively, are then 0.90 and 0.66 solar mass.

The star known as 61 Cygni was the first whose distance was measured (via parallax by the German astronomer Friedrich W. Bessel in the mid-19th century). Visually, 61 Cygni is a double star separated by 83.2 astronomical units. Its members move around one another with a period of 653 years. It was among the first stellar systems thought to contain a potential planet, although this has not been confirmed and is now considered unlikely. Nevertheless, since the 1990s a variety of discovery techniques have confirmed the existence of thousands of planets orbiting other stars.

Spectroscopic Binaries

Spectroscopic binary stars are found from observations of radial velocity. At least the brighter member of such a binary can be seen to have a continuously changing periodic velocity that alters the wavelengths of its spectral lines in a rhythmic way; the velocity curve repeats itself exactly from one cycle to the next, and the motion can be interpreted as orbital motion. In some cases, rhythmic changes in the lines of both members can be measured. Unlike visual binaries, the semimajor axes or the individual masses cannot be found for most spectroscopic binaries, since the angle between the orbit plane and the plane of the sky cannot be determined. If spectra from both members are observed, mass ratios can be found. If one spectrum alone is observed, only a quantity called the mass function can be derived, from which is calculated a lower limit to the stellar masses. If a spectroscopic binary is also observed to be an eclipsing system, the inclination of the orbit and often the values of the individual masses can be ascertained.

Eclipsing Binaries

An eclipsing binary consists of two close stars moving in an orbit so placed in space in relation to Earth that the light of one can at times be hidden behind the other. Depending on the orientation of the orbit and sizes of the stars, the eclipses can be total or annular (in the latter, a ring of one star shows behind the other at the maximum of the eclipse) or both eclipses can be partial. The best known example of an eclipsing binary is Algol (Beta Persei), which has a period (interval between eclipses) of 2.9 days. The brighter (B8-type) star contributes about 92 percent of the light of the system, and the eclipsed star provides less than 8 percent. The system contains a third star that is not eclipsed. Some 20 eclipsing binaries are visible to the naked eye.

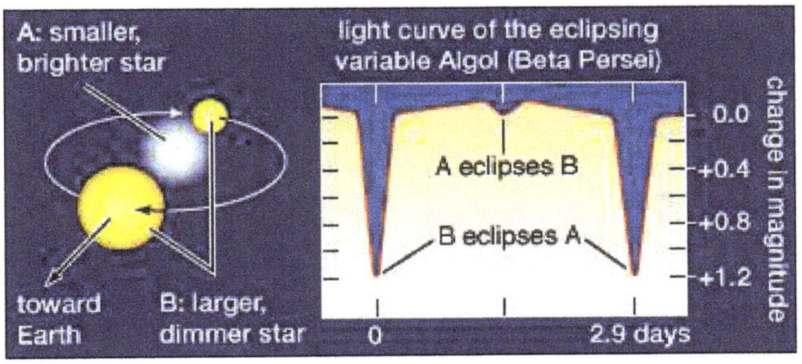

Light curve of Algol (Beta Persei), an eclipsing variable, or eclipsing binary, star system.

The relative brightness of the system, in the figure above, is plotted against time. A sharp dip occurs every 2.9 days when the fainter component star eclipses the brighter one, a shallower dip when the brighter star eclipses the fainter one.

The light curve for an eclipsing binary displays magnitude measurements for the system over a complete light cycle. The light of the variable star is usually compared with that of a nearby (comparison) star thought to be fixed in brightness. Often, a deep, or primary, minimum is produced when the component having the higher surface brightness is eclipsed. It represents the total eclipse and is characterized by a flat bottom. A shallower secondary eclipse occurs when the brighter component passes in front of the other; it corresponds to an annular eclipse (or transit). In a partial eclipse neither star is ever completely hidden, and the light changes continuously during an eclipse.

The shape of the light curve during an eclipse gives the ratio of the radii of the two stars and also one radius in terms of the size of the orbit, the ratio of luminosities, and the inclination of the orbital plane to the plane of the sky.

If radial-velocity curves are also available—i.e., if the binary is spectroscopic as well as eclipsing—additional information can be obtained. When both velocity curves are observable, the size of the orbit as well as the sizes, masses, and densities of the stars can be calculated. Furthermore, if the distance of the system is measurable, the brightness temperatures of the individual stars can be estimated from their luminosities and radii. All of these procedures have been carried out for the faint binary Castor C (two red-dwarf components of the six-member Castor multiple star system) and for the bright B-type star Mu Scorpii.

Close stars may reflect each other's light noticeably. If a small, high-temperature star is paired with a larger object of low surface brightness and if the distance between the stars is small, the part of the cool star facing the hotter one is substantially brightened by it. Just before (and just after) secondary eclipse, this illuminated hemisphere is pointed toward the observer, and the total light of the system is at a maximum.

The properties of stars derived from eclipsing binary systems are not necessarily applicable to isolated single stars. Systems in which a smaller, hotter star is accompanied by a larger, cooler object are easier to detect than are systems that contain, for example, two main-sequence stars. In such an unequal system, at least the cooler star has certainly been affected by evolutionary changes, and probably so has the brighter one. The evolutionary development of two stars near one another does not exactly parallel that of two well-separated or isolated ones.

Eclipsing binaries include combinations of a variety of stars ranging from white dwarfs to huge supergiants (e.g., VV Cephei), which would engulf Jupiter and all the inner planets of the solar system if placed at the position of the Sun.

Some members of eclipsing binaries are intrinsic variables, stars whose energy output fluctuates with time. In many such systems, large clouds of ionized gas swirl between the stellar members. In others, such as Castor C, at least one of the faint M-type dwarf components might be a flare star, one in which the brightness can unpredictably and suddenly increase to many times its normal value.

Binaries and Extrasolar Planetary Systems

Near the Sun, most stars are members of binaries, and many of the nearest single stars are suspected of having companions. Although some binary members are separated by hundreds of astronomical units and others are contact binaries (stars close enough for material to pass between them), binary systems are most frequently built on the same scale as that of the solar system—namely, on the order of about 10 astronomical units. The division in mass between two components of a binary seems to be nearly random. A mass ratio as small as about 1:20 could occur about 5 percent of the time, and under these circumstances a planetary system comparable to the solar system is able to form.

The formation of double and multiple stars on the one hand and that of planetary systems on the other seem to be different facets of the same process. Planets are probably produced as a natural by-product of star formation. Only a small fraction of the original nebula matter is likely to be retained in planets, since much of the mass and angular momentum is swept out of the system. Conceivably, as many as 100 million stars could have bona fide planets in the Milky Way Galaxy.

Individual planets around other stars—i.e., extrasolar planets—are very difficult to observe directly because a star is always much brighter than its attendant planet. Jupiter, for example, would be only one-billionth as bright as the Sun and appear so close to it as to be undetectable from even the nearest star. If candidate stars are treated as possible spectroscopic binaries, however, then one may look for a periodic change in the star's radial velocity caused by a planet swinging around it. The effect is very small: even Jupiter would cause a change in the apparent radial velocity of the Sun of only about 10 metres (33 feet) per second spread over Jupiter's orbital period of about 12 years at best. Current techniques using very large telescopes to study fairly bright stars can measure radial velocities with a precision of about a metre per second, provided that the star has very sharp spectral lines, such as is observed for Sun-like stars and stars of types K and M. This means that at present the radial-velocity method normally can detect planets like Earth around M-type stars. Moreover, the closer the planet is to its parent star, the greater and quicker the velocity swing, so that detection of giant planets close to a star is favoured over planets farther out. Finally, even when a planet is detected, the usual spectroscopic binary problem of not knowing the angle between the orbit plane and that of the sky allows only a minimum mass to be assigned to the planet.

One exception to this last problem is HD 209458, a seventh-magnitude G0 V star about 150 light-years away with a planetary object orbiting it every 3.5 days. Soon after the companion, HD 209458b, was discovered in 1999 by its effect on the star's radial velocity, it also was found to be eclipsing the star, meaning that its orbit is oriented almost edge-on toward Earth. This type of eclipse is called a transit, and this method has been used, most notably by the Kepler satellite, to find thousands of extrasolar planets. Some of these planets are roughly the size of Earth and can be found in their star's habitable zone, the distance from a star where liquid water, and thus possibly life, can survive on the surface.

Extrasolar planet Kepler-16b. The first planet found to orbit two stars, Kepler-16b.

This transit of HD 209458b, as well as observations of spectral lines in the planet's atmosphere, allowed determination of the planet's mass and radius—0.64 and 1.38 times those of Jupiter, respectively. These numbers imply that the planet is even more of a giant than Jupiter itself. What was unexpected is its proximity to the parent star—more than 100 times closer than Jupiter is to the Sun, raising the question of how a giant gaseous planet that closes can survive the star's radiation. The fact that many other extrasolar planets have been found to have orbital periods measured in days rather than years, and thus to be very close to their parent stars, suggests that the HD 209458 case is not unusual. There are also some confirmed cases of planets around supernova remnants called pulsars, although whether the planets preceded the supernova explosions that produced the pulsars or were acquired afterward remains to be determined.

The first extrasolar planets were discovered in 1992. Thousands of extrasolar planets are known, with more such discoveries being added regularly.

In addition to the growing evidence for existence of extrasolar planets, space-based observatories designed to detect infrared radiation have found many young stars (including Vega, Fomalhaut, and Beta Pictoris) to have disks of warm matter orbiting them. This matter is composed of myriad particles mostly about the size of sand grains and might be taking part in the first stage of planetary formation.

The infrared emission from the young star Fomalhaut and the dust belt surrounding it, as recorded with the European Space Agency's satellite observatory Herschel.

Mass Extremes

The mass of most stars lies within the range of 0.3 to 3 solar masses. The star with the largest mass determined to date is R136a1, a giant of about 265 solar masses that had as much as 320 solar masses when it was formed. There is a theoretical upper limit to the masses of nuclear-burning stars (the Eddington limit), which limits stars to no more than a few hundred solar masses. On the low mass side, most stars seem to have at least 0.1 solar mass. The theoretical lower mass limit for an ordinary star is about 0.075 solar mass, for below this value an object cannot attain a central temperature high enough to enable it to shine by nuclear energy. Instead, it may produce a much lower level of energy by gravitational shrinkage. If its mass is not much below the critical 0.075 solar mass value, it will appear as a very cool, dim star known as a brown dwarf. Its evolution is simply to continue cooling toward eventual extinction. At still somewhat lower masses, the object would be a giant planet. Jupiter, with a mass roughly 0.001 that of the Sun, is just such an object, emitting a very low level of energy (apart from reflected sunlight) that is derived from gravitational shrinkage.

Brown dwarfs were late to be discovered, the first unambiguous identification having been made in 1995. It is estimated, however, that hundreds must exist in the solar neighbourhood. An extension of the spectral sequence for objects cooler than M-type stars has been constructed, using L for warmer brown dwarfs, T for cooler ones, and Y for the coolest. The presence of methane in the T brown dwarfs and of ammonia in the Y brown dwarfs emphasizes their similarity to giant planets.

The brown dwarf 2MASSWJ 1207334−393254 (centre) as seen in a photo taken by the Very Large Telescope at the European Southern Observatory, Cerro Paranal, Chile. The brown dwarf has a mass 25 times that of Jupiter and a surface temperature of 2,400 K. Orbiting the brown dwarf at a distance of 8.3 billion km (5.2 billion miles) is a planet (lower left) that has a mass five times that of Jupiter and a surface temperature of 1,250 K.

Stellar Radii

Angular sizes of bright red giant and supergiant stars were first measured directly during the 1920s, using the principle of interference of light. Only bright stars with large angular size can be measured by this method. Provided the distance to the star is known, the physical radius can be determined.

Eclipsing binaries also provide extensive data on stellar dimensions. The timing of eclipses provides the angular size of any occulting object, and so analyzing the light curves of eclipsing binaries can be a useful means of determining the dimensions of either dwarf or giant stars. Members of close binary systems, however, are sometimes subject to evolutionary effects, mass exchange, and other disturbances that change the details of their spectra.

A more recent method, called speckle interferometry, has been developed to reproduce the true disks of red supergiant stars and to resolve spectroscopic binaries such as Capella. The speckle phenomenon is a rapidly changing interference-diffraction effect seen in a highly magnified diffraction image of a star observed with a large telescope.

If the absolute magnitude of a star and its temperature are known, its size can be computed. The temperature determines the rate at which energy is emitted by each unit of area, and the total luminosity gives the total power output. Thus, the surface area of the star and, from it, the radius of the object can be estimated. This is the only way available for estimating the dimensions of white dwarf stars. The chief uncertainty lies in choosing the temperature that represents the rate of energy emission.

Average Stellar Values

Main-sequence stars range from very luminous objects to faint M-type dwarf stars, and they vary considerably in their surface temperatures, their bolometric (total) luminosities, and their radii. Moreover, for stars of a given mass, a fair spread in radius, luminosity, surface temperature, and spectral type may exist. This spread is produced by stellar evolutionary effects and tends to broaden the main sequence. Masses are obtained from visual and eclipsing binary systems observed spectroscopically. Radii are found from eclipsing binary systems, from direct measurements in a few favourable cases, by calculations, and from absolute visual magnitudes and temperatures.

Average values for radius, bolometric luminosity, and mass are meaningful only for dwarf stars. Giant and subgiant stars all show large ranges in radius for a given mass. Conversely, giant stars of very nearly the same radius, surface temperature, and luminosity can have appreciably different masses.

Stellar Statistics

Some of the most important generalizations concerning the nature and evolution of stars can be derived from correlations between observable properties and from certain statistical results. One of the most important of these correlations concerns temperature and luminosity—or, equivalently, colour and magnitude.

Hertzsprung-Russell Diagram

When the absolute magnitudes of stars, or their intrinsic luminosities on a logarithmic scale, are plotted in a diagram against temperature or, equivalently, against the spectral types, the stars do not fall at random on the diagram but tend to congregate in certain restricted domains. Such a plot is usually called a Hertzsprung-Russell diagram, named for the early 20th-century astronomers Ejnar Hertzsprung of Denmark and Henry Norris Russell of the United States, who independently discovered the relations shown in it. As is seen in the diagram, most of the congregated stars are dwarfs lying closely around a diagonal line called the main sequence. These stars range from hot, O- and B-type, blue objects at least 10,000 times brighter than the Sun down through white A-type stars such as Sirius to orange K-type stars such as Epsilon Eridani and finally to M-type red dwarfs thousands of times fainter than the Sun. The sequence is continuous; the luminosities fall off smoothly with decreasing surface temperature; the masses and radii decrease but at a much slower rate; and the stellar densities gradually increase.

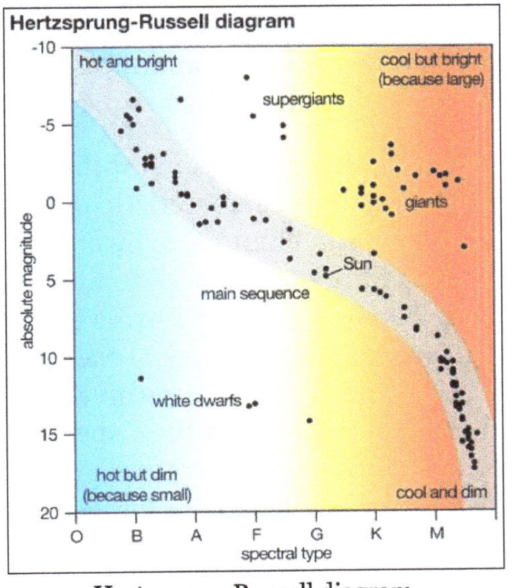

Hertzprung-Russell diagram.

The second group of stars to be recognized was a group of giants—such objects as Capella, Arcturus, and Aldebaran—which are yellow, orange, or red stars about 100 times as bright as the Sun and have radii on the order of 10–30 million km (about 6–20 million miles, or 15–40 times as large as the Sun). The giants lie above the main sequence in the upper right portion of the diagram. The category of supergiants includes stars of all spectral types; these stars show a large spread in intrinsic brightness, and some even approach absolute magnitudes of −7 or −8. A few red supergiants, such as the variable star VV Cephei, exceed in size the orbit of Jupiter or even that of Saturn, although most of them are smaller. Supergiants are short-lived and rare objects, but they can be seen at great distances because of their tremendous luminosity.

Subgiants are stars that are redder and larger than main-sequence stars of the same luminosity. Many of the best-known examples are found in close binary systems where conditions favour their detection.

The white dwarf domain lies about 10 magnitudes below the main sequence. These stars are in the last stages of their evolution.

The spectrum-luminosity diagram has numerous gaps. Few stars exist above the white dwarfs and to the left of the main sequence. The giants are separated from the main sequence by a gap named for Hertzsprung, who in 1911 became the first to recognize the difference between main-sequence and giant stars. The actual concentration of stars differs considerably in different parts of the diagram. Highly luminous stars are rare, whereas those of low luminosity are very numerous.

The spectrum-luminosity diagram applies to the stars in the galactic spiral arm in the neighbourhood of the Sun and represents what would be obtained if a composite Hertzsprung-Russell diagram were constructed combining data for a large number of the star groups called open (or galactic) star clusters, as, for example, the double cluster h and χ Persei, the Pleiades, the Coma cluster, and the Hyades. It includes very young stars, a few million years old, as well as ancient stars perhaps as old as 10 billion years.

By contrast, another Hertzsprung-Russell diagram exhibits the type of temperature-luminosity, or colour-magnitude, relation characteristic of stars in globular clusters, in the central bulge of the Galaxy, and in elliptical external galaxies—namely, of the so-called stellar Population II. (In addition to these oldest objects, Population II includes other very old stars that occur between the spiral arms of the Galaxy and at some distance above and below the galactic plane). Because these systems are very remote from the observer, the stars are faint, and their spectra can be observed only with difficulty. As a consequence, their colours rather than their spectra must be measured. Since the colours are closely related to surface temperature and therefore to spectral types, equivalent spectral types may be used, but it is stellar colours, not spectral types, that are observed in this instance.

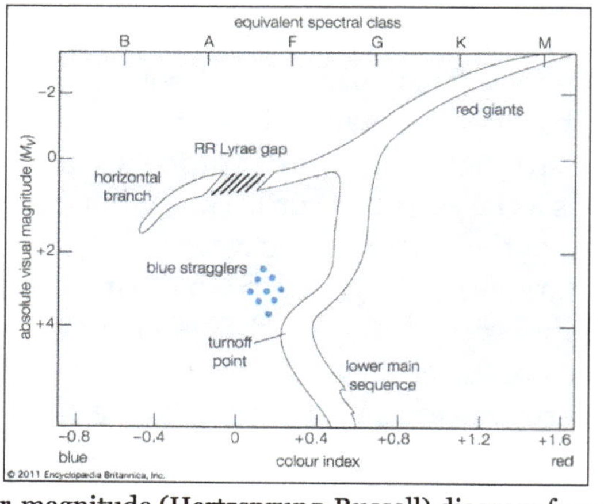

Colour-magnitude (Hertzsprung-Russell) diagram for an old globular cluster made up of Population II stars.

The differences between the two Hertzsprung-Russell diagrams are striking. In the second there are no supergiants, and, instead of a domain at an absolute magnitude of about 0, the giant stars form a branch that starts high and to the right at about −3.5 for very red stars and flows in a continuous sequence until it reaches an absolute magnitude of about 0. At that point the giant branch splits: a main band of stars, all about the same colour, proceeds downward (i.e., to fainter stars) to a magnitude of about +3 and then connects to the main sequence at about +4 by way of a narrow band. The main sequence of Population II stars extends downward to fainter, redder stars in much the same way as in the spiral-arm Population I stars. (Population I is the name given to the stars found within the spiral arms of the Milky Way system and other galaxies of the same type. Containing stars of all ages, from those in the process of formation to defunct white dwarfs, Population I is nonetheless always associated with the gas and dust of the interstellar medium). The main sequence ends at about spectral type G, however, and does not extend up through the A, B, and O spectral types, though occasionally a few such stars, blue stragglers, are found in the region normally occupied by the main sequence. The blue stragglers are likely red stars that have gained mass from another star through stellar collision or mass transfer.

The other band of stars formed from the split of the giant branch is the "horizontal branch," which falls near magnitude +0.6 and fills the aforementioned Hertzsprung gap, extending to increasingly blue stars beyond the RR Lyrae stars, which are indicated by the crosshatched area in the diagram. Among these blue hot stars are found novas and the nuclei of planetary nebulas, the latter so called because their photographic image resembles that of a distant planet. Not all globular clusters show identical colour-magnitude diagrams, which may be due to differences in the cluster ages or other factors.

Estimates of Stellar Ages

The shapes of the colour-magnitude diagrams permit estimates of globular-cluster ages. The point at which stars move away from the main sequence is called the turnoff point, For example, in a cluster where stars more massive than about 1.3 solar masses have evolved away from the main sequence at a point just above the position occupied by the Sun, the time required for such a star to exhaust the hydrogen in its core is about 5–6 billion years, and the cluster must be at least as old. More ancient clusters have been identified. In the Galaxy, globular clusters are all very ancient objects, having ages within a few billion years of the average of 11 billion years. In the Magellanic Clouds, however, clusters exist that resemble globular ones, but they contain numerous blue stars and therefore must be relatively young.

Small Magellanic Cloud. Infant stars in the Small Magellanic Cloud.

Open clusters in the spiral arms of the Galaxy—extreme Population I—tell a somewhat different story. A colour-magnitude diagram can be plotted for a number of different open clusters—for example, the double cluster h and χ Persei, the Pleiades, Praesepe, and M67—with the main feature distinguishing the clusters being their ages. The young cluster h and χ Persei, which is a few million years old, contains stars ranging widely in luminosity. Some stars have already evolved into the supergiant stage (in such a diagram the top of the main sequence is bent over). The stars of luminosity 10,000 times greater than that of the Sun have already largely depleted the hydrogen in their cores and are leaving the main sequence.

The brightest stars of the Pleiades cluster, aged about 100 million years, have begun to leave the main sequence and are approaching the critical phase when they will have exhausted all the hydrogen in their cores. There are no giants in the Pleiades. Presumably, the cluster contained no stars as massive as some of those found in h and χ Persei.

Pleiades: Bright nebulosity in the Pleiades (M45, NGC 1432), distance 490 light-years. Cluster stars provide the light, and surrounding clouds of dust reflect and scatter the rays from the stars.

The cluster known as Praesepe, or the Beehive, at an age of 790 million years, is older than the Pleiades. All stars much more luminous than the first magnitude have begun to leave the main sequence; there are some giants. The Hyades, about 620 million years old, displays a similar colour-magnitude array. These clusters contain a number of white dwarfs, indicating that the initially most luminous stars have already run the gamut of evolution. In a very old cluster such as M67, which is 4.5 billion years old, all of the bright main-sequence stars have disappeared.

The colour-magnitude diagrams for globular and open clusters differ quantitatively because the latter show a wider range of ages and differ in chemical composition. Most globular clusters have smaller metal-to-hydrogen ratios than do open clusters or the Sun. The gaps between the red giants and the blue main-sequence stars of the open clusters (Population I) often contain unstable stars such as variables. The Cepheid variable stars, for instance, fall in these gaps.

Cepheid variables: Cepheid variables, as seen by the Hubble Space Telescope.

The giant stars of the Praesepe cluster are comparable to the brightest stars in M67. The M67 giants have evolved from the main sequence near an absolute magnitude of +3.5, whereas the Praesepe giants must have masses about twice as great as those of the M67 giants. Giant stars of the same luminosity may therefore have appreciably different masses.

Numbers of Stars vs. Luminosity

Of great statistical interest is the relationship between the luminosities of the stars and their frequency of occurrence. The naked-eye stars are nearly all intrinsically brighter than the Sun, but the opposite is true for the known stars within 20 light-years of the Sun. The bright stars are easily seen at great distances; the faint ones can be detected only if they are close.

The luminosity function (the number of stars with a specific luminosity) depends on population type. The luminosity function for pure Population II differs substantially from that for pure Population I. There is a small peak near absolute magnitude +0.6, corresponding to the horizontal branch for Population II, and no stars as bright as absolute magnitude −5. The luminosity function for pure Population I is evaluated best from open star clusters, the stars in such a cluster being at about the same distance. The neighbourhood of the Sun includes examples of both Populations I and II.

Mass-luminosity Correlations

A plot of mass against bolometric luminosity for visual binaries for which good parallaxes and masses are available shows that for stars with masses comparable to that of the Sun the luminosity, L, varies as a power, $3 + \beta$, of the mass M. This relation can be expressed as $L = (M)^{3+\beta}$. The power differs for substantially fainter or much brighter stars.

This mass-luminosity correlation applies only to unevolved main-sequence stars. It fails for giants and supergiants and for the subgiant (dimmer) components of eclipsing binaries, all of which have changed considerably during their lifetimes. It does not apply to any stars in a globular cluster not on the main sequence, or to white dwarfs that are abnormally faint for their masses.

The mass-luminosity correlation, predicted theoretically in the early 20th century by the English astronomer Arthur Eddington, is a general relationship that holds for all stars having essentially the same internal density and temperature distributions—i.e., for what are termed the same stellar models.

Variable Stars

Many stars are variable. Some are geometric variables, as in the eclipsing binaries considered earlier. Others are intrinsically variable—i.e., their total energy output fluctuates with time. Such intrinsic variable stars are dealt with in this topic.

A fair number of stars are intrinsically variable. Some objects of this type were found by accident, but many were detected as a result of carefully planned searches. Variable stars are important in astronomy for several reasons. They usually appear to be stars at critical or short-lived phases of their evolution, and detailed studies of their light and spectral characteristics, spatial distribution, and association with other types of stars may provide valuable clues to the life histories of various classes of stars. Certain kinds of variable stars, such as Cepheids (periodic variables) and novas and supernovas (explosive variables), are extremely important in that they make it possible to establish the distances of remote stellar systems beyond the Galaxy. If the intrinsic luminosity of a recognizable variable is known and this kind of variable star can be found in a distant stellar system, the distance of the latter can be estimated from a measurement of apparent and absolute magnitudes, provided that the interstellar absorption is also known.

Classification

Variables are often classified as behaving like a prototype star, and the entire class is then named for this star—e.g., RR Lyrae stars are those whose variability follows the pattern of the star RR Lyrae. The most important classes of intrinsically variable stars are the following:

- Pulsating variables—stars whose variations in light and colour are thought to arise primarily from stellar pulsations. These include Beta Canis Majoris stars, RR Lyrae stars, and Delta Scuti stars, all with short regular periods of less than a day; Cepheids, with periods between 1 and 100 days; and long-period variables, semiregular variables, and irregular red variables, usually with unstable periods of hundreds of days.
- Explosive, or catastrophic, variables—stars in which the variations are produced by the wrenching away of part of the star, usually the outer layers, in some explosive process. They include SS Cygni or U Geminorum stars, novas, and supernovas (the last of which are enormous explosions involving most of the matter in a star.

- Miscellaneous and special types of variables—R Coronae Borealis stars, T Tauri stars, flare stars, pulsars (neutron stars), spectrum and magnetic variables, X-ray variable stars, and radio variable stars.

Pulsating Stars

An impressive body of evidence indicates that stellar pulsations can account for the variability of Cepheids, long-period variables, semiregular variables, Beta Canis Majoris stars, and even the irregular red variables. Of this group, the Cepheid variables have been studied in greatest detail, both theoretically and observationally. These stars are regular in their behaviour; some repeat their light curves with great faithfulness from one cycle to the next over periods of many years.

Much confusion existed in the study of Cepheids until it was recognized that different types of Cepheids are associated with different groups, or population types, of stars. Cepheids belonging to the spiral-arm Population I (or Type I Cepheids) are characterized by regularity in their behaviour. They show continuous velocity curves indicative of regular pulsation. They exhibit a relation between period and luminosity in the sense that the longer the period of the star, the greater its intrinsic brightness. This period-luminosity relationship has been used to establish the distances of remote stellar systems.

Cepheids with different properties are found in Population II, away from the Milky Way, in globular clusters. These Type II Cepheids are bluer than classic Population I Cepheids having the same period, and their light curves have different shapes. Studies of the light and velocity curves indicate that shells of gas are ejected from the stars as discontinuous layers that later fall back toward the surface. These stars exhibit a relation between period and luminosity different from that for Population I Cepheids, and thus the distance of a Cepheid in a remote stellar system can be determined only if its population type is known.

Closely associated with Population II Cepheids are the cluster-type, or RR Lyrae, variables. Many of these stars are found in clusters, but some, such as the prototype RR Lyrae, occur far from any cluster or the central galactic bulge. Their periods are less than a day, and there is no correlation between period and luminosity. Their absolute magnitudes are about 0.6, but somewhat dependent on metal abundance. They are thus about 50 times as bright as the Sun and so are useful for determining the distance of star clusters and some of the nearer external galaxies, their short periods permitting them to be detected readily.

Long-period variable stars also probably owe their variations to pulsations. Here the situation is complicated by the vast extent of their atmospheres, so that radiation originating at very different depths in the star is observed at the same time. At certain phases of the variations, bright hydrogen lines are observed overlaid with titanium oxide absorption. The explanation is an outward-moving layer of hot, recombining gas, whose radiation is absorbed by strata of cool gases. These stars are all cool red giants and supergiants of spectral types M (normal composition), R and N (carbon-rich), or S (heavy-metal-rich). The range in visual brightness during a pulsation can be 100-fold, but the range in total energy output is much less, because at very low stellar temperatures (1,500–3,000 K) most of the energy is radiated in the infrared as heat rather than as light.

Unlike the light curves of classic Cepheids, the light curves of these red variables show considerable variations from one cycle to another. The visual magnitude of the variable star Mira Ceti (Omicron Ceti) is normally about 9–9.5 at minimum light, but at maximum it may lie between 5 and 2. Time intervals between maxima often vary considerably. In such cool objects, a very small change in temperature can produce a huge change in the output of visible radiation. At the low temperatures of the red variables, compounds and probably solid particles are formed copiously, so that the visible light may be profoundly affected by a slight change in physical conditions. Random fluctuations from cycle to cycle, which would produce negligible effects in a hotter star, produce marked light changes in a long-period variable.

Long-period variables appear to fall into two groups; those with periods of roughly 200 days tend to be associated with Population II, and those of periods of about a year belong to Population I.

Red semiregular variables such as the RV Tauri stars show complex light and spectral changes. They do not repeat themselves from one cycle to the next; their behaviour suggests a simultaneous operation of two or more modes of oscillation. Betelgeuse is an example of an irregular red variable. In these stars the free period of oscillation does not coincide with the periodicity of the driving mechanism.

Finally, among the various types of pulsating variable stars, the Beta Canis Major is variables are high-temperature stars (spectral type B) that often show complicated variations in spectral-line shapes and intensities, velocity curves, and light. In many cases, they have two periods of variation so similar in duration that complex interference or beat phenomena are observed, both in radial velocities and in the shapes of spectral lines.

A large body of evidence suggests that all members of this first class of variable stars owe their variability to pulsation. The pulsation theory was first proposed as a possible explanation as early as 1879, was applied to Cepheids in 1914, and was further developed by Arthur Eddington in 1917–18. Eddington found that if stars have roughly the same kind of internal structure, then the period multiplied by the square root of the density equals a constant that depends on the internal structure.

The Eddington theory, though a good approximation, encountered some severe difficulties that have been met through modifications. If the entire star pulsated in synchronism, it should be brightest when compressed and smaller while faintest when expanded and at its largest. The radial velocity should be zero at both maximum and minimum light. Observations contradict these predictions. When the star pulsates, all parts of the main body move in synchronism, but the outer observable strata fall out of step or lag behind the pulsation of the inner regions. Pulsations involve only the outer part of a star; the core, where energy is generated by thermonuclear reactions, is unaffected.

Careful measurements of the average magnitudes and colours of RR Lyrae stars in the globular cluster M3 showed that all these stars fell within a narrow range of luminosity and colour (or surface temperature) or, equivalently, luminosity and radius. Also, every star falling in this narrow range of brightness and size was an RR Lyrae variable. Subsequent work has indicated that similar considerations apply to most classic Cepheids. Variability is thus a characteristic of any star whose evolution carries it to a certain size and luminosity, although the amplitude of the variability can vary dramatically.

In the pulsation theory as now developed, the light and velocity changes of Cepheids can be interpreted not only qualitatively but also quantitatively. The light curves of Cepheids, for example, have been precisely predicted by the theory. Stellar pulsation, like other rhythmic actions, may give rise to harmonic phenomena wherein beats reinforce or interfere with one another. Beat and interference phenomena then complicate the light and velocity changes. The RR Lyrae stars supply some of the best examples, but semiregular variables such as the RV Tauri stars or most Delta Scuti stars evidently vibrate simultaneously with two or more periods.

Explosive Variables

The evolution of a member of a close double-star system can be markedly affected by the presence of its companion. As the stars age, the more massive one swells up more quickly as it moves away from the main sequence. It becomes so large that its outer envelope falls under the gravitational influence of the smaller star. Matter is continuously fed from the more rapidly evolving star to the less massive one, which still remains on the main sequence. U Cephei is a classic example of such a system for which spectroscopic evidence shows streams of gas flowing from the more highly evolved star to the hotter companion, which is now the more massive of the two. Eventually, the latter will also leave the main sequence and become a giant star, only to lose its outer envelope to the companion, which by that time may have reached the white dwarf stage.

Novas appear to be binary stars that have evolved from contact binaries of the W Ursae Majoris type, which are pairs of stars apparently similar to the Sun in size but revolving around one another while almost touching. One member may have reached the white dwarf stage. Matter fed to it from its distended companion appears to produce instabilities that result in violent explosions or nova outbursts. The time interval between outbursts can range from a few score years to hundreds of thousands of years.

In ordinary novas the explosion seems to involve only the outer layers, as the star later returns to its former brightness; in supernovas the explosion is catastrophic. Normally, novas are small blue stars much fainter than the Sun, though very much hotter.

When an outburst occurs, the star can brighten very rapidly, by 10 magnitudes or more in a few hours. Thereafter it fades; the rate of fading is connected with the brightness of the nova. The brightest novas, which reach absolute magnitudes of about −10, fade most rapidly, whereas a typical slow nova, which reaches an absolute magnitude of −5, can take 10 or 20 times as long to decline in brightness. This property, when calibrated as the absolute magnitude at maximum brightness versus the time taken to decline by two magnitudes, allows novas to be used as distance indicators for nearby galaxies. The changes in light are accompanied by pronounced spectroscopic changes that can be interpreted as arising from alterations in an ejected shell that dissipates slowly in space. In its earliest phases, the expanding shell is opaque. As its area grows, with a surface temperature near 7,000 K, the nova brightens rapidly. Then, near maximum light, the shell becomes transparent, and its total brightness plummets rapidly, causing the nova to dim.

The mass of the shell is thought to be rather small, about 10–100 times the mass of Earth. Only the outer layers of the star seem to be affected; the main mass settles down after the outburst into a state much as before until a new outburst occurs. The existence of repeating novas, such as the star T Coronae Borealis, suggests that perhaps all novas repeat at intervals ranging up to thousands or perhaps millions of years; and probably, the larger the explosion, the longer the interval. There is strong evidence that novas are components of close double stars and, in particular that they have evolved from the most common kind of eclipsing binaries, those of the W Ursae Majoris type.

Stars of the SS Cygni type, also known as dwarf novas, undergo novalike outbursts but of a much smaller amplitude. The intervals between outbursts are a few months to a year. Such variables are close binaries. The development of this particular type may be possible only in close binary systems.

There are two major types of supernovas, designated Type I (or SNe I) and Type II (or SNe II). They can be distinguished by the fact that Type II supernovas have hydrogen features in their spectra, while Type I supernovas do not. Type II supernovas arise from the collapse of a single star more massive than about eight solar masses, resulting in either a neutron star or a black hole. There are three classes of Type I supernovas: Ia, Ib, and Ic. Type Ia supernovas are believed to originate in a binary system containing a white dwarf, rather like the case of ordinary novas. Unlike the latter, however, in which only the outer layers of the white dwarf seem to be affected, in a Type I supernova the white dwarf is probably completely destroyed; the details are not yet fully understood. Certainly, a supernova's energy output is enormously greater than that of an ordinary nova. Type Ib and Ic supernovas are like Type II in that they are each the core collapse of a massive star. However, a Type II supernova retains its hydrogen envelope, whereas the Type Ib and Ic supernovas do not, thus leading to the different hydrogen features in their spectra. Type Ib retains a helium shell and so has a spectrum rich in helium lines; Type Ic does not retain the hydrogen or helium shell.

Supernova 1987A in the Large Magellanic Cloud.

Empirical evidence indicates that in a Type Ia supernova the absolute magnitude at maximum light can be determined by a combination of data derived from the rate of dimming after maximum, the shape of the light curve, and certain colour measurements. A comparison of the absolute and apparent magnitudes of maximum light in turn allows the distance of the supernova to be found. This is a matter of great usefulness because Type Ia supernovas at maximum light are the most luminous "standard candles" available for determining distances to external galaxies and thus can be observed in more distant galaxies than any other kind of standard candle. In 1999, application of this technique led to the totally unexpected discovery that the expansion of the universe is accelerating rather than slowing down. This acceleration is caused by dark energy, a gravitationally repulsive force that is the dominant component (73 percent) of the universe's mass-energy.

Supernova: Three distant Type Ia supernovas, as observed by the Hubble Space Telescope. The bottom images are details of the upper wide views. The supernovas at left and centre occurred about five billion years ago, the right seven billion years ago.

A peculiar explosive variable with no known counterpart is Eta Carinae (NGC 3372), which appears in telescopes on Earth as a fuzzy red "star" slightly less than two seconds of arc in diameter. Surrounding it is a shell of gas and dust shaped roughly like an hourglass divided by a thin disk. First observed as a star of about the fourth magnitude in 1677, it brightened irregularly, undergoing an outburst in 1843, when it became for a few years the second brightest star in the sky. Thereafter it slowly faded, becoming too faint for the unaided eye around the turn of the 20th century. The fading was due, at least in part, to obscuration by dust emitted in the earlier eruption. The star remained near seventh magnitude with irregular variations for most of the 20th century, but it began brightening again by one or two tenths of a magnitude per year in the mid-1990s. In 2005 astronomers found that Eta Carinae is, in fact, a binary star system with an orbital period of 5.52 years. Its A component has a temperature of about 15,000 K; its B component, about 35,000 K. Eta Carinae is considered to be one of a small class of stars known as luminous blue variables. Its luminosity has been estimated as five million times that of the Sun. Flaring events producing not only visible effects but also X-ray, ultraviolet, and radio-wave effects have been observed.

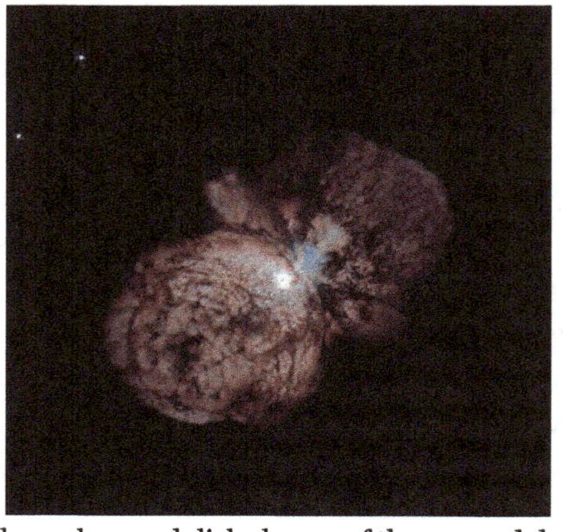

Eta Carinae: The hourglass and disk shapes of the gas and dust clouds, as well as various smaller details, are visible in this enhanced, computer-manipulated composite of eight images taken by the Hubble Space Telescope.

Probably all variable stars represent more or less ephemeral phases in the evolution of a star. Aside from catastrophic events of the kind that produce a supernova, some phases of stellar variability might be of such brief duration as to permit recognizable changes during an interval of 50–100 years. Other stages may require many thousands of years. For example, the period of Delta Cephei, the prototype star of the Cepheid variables, has barely changed by a detectable amount since its variability was discovered in 1784.

Peculiar Variables

R Coronae Borealis variables are giant stars of about the Sun's temperature whose atmospheres are characterized by excessive quantities of carbon and very little hydrogen. The brightness of such a star remains constant until the star suddenly dims by several magnitudes and then slowly recovers its original brightness. (The star's colour remains the same during the changes in brightness). The dimmings occur in a random fashion and seem to be due to the huge concentrations of carbon. At times the carbon vapour literally condenses into soot, and the star is hidden until the smog blanket is evaporated. Similar veiling may sometimes occur in other types of low-temperature stars, particularly in long-period variables.

Flare stars are cool dwarfs (spectral type M) that display flares apparently very much like, but much more intense than, those of the Sun. In fact, the flares are sometimes so bright that they overwhelm the normal light of the star. Solar flares are associated with copious emission of radio waves, and simultaneous optical and radio-wave events appear to have been found in the stars UV Ceti, YZ Canis Minoris, and V371 Orionis.

Spectrum and magnetic variables, mostly of spectral type A, show only small amplitudes of light variation but often pronounced spectroscopic changes. Their spectra typically show strong lines of metals such as manganese, titanium, iron, chromium, and the lanthanides (also called rare earths), which vary periodically in intensity. These stars have strong magnetic fields, typically from a few hundred to a few thousand gauss. One star, HD 215441, has a field on the order of 30,000 gauss. (Earth's magnetic field has an average strength of about 0.5 gauss). Not all magnetic stars are known to be variable in light, but such objects do seem to have variable magnetic fields. The best interpretation is that these stars are rotating about an inclined axis. As with Earth, the magnetic and rotation axes do not coincide. Different ions are concentrated in different areas (e.g., chromium in one area and the lanthanides in another).

The Sun is an emitter of radio waves, but, with present techniques, its radio emission could only just be detected from several parsecs away. Most discrete radio-frequency sources have turned out to be objects such as old supernovas, radio galaxies, or quasars, though well-recognized radio stars also have been recorded on occasion. These include flare stars, red supergiants such as Betelgeuse, the high-temperature dwarf companion to the red supergiant Antares, and the shells ejected from Nova Serpentis 1970 and Nova Delphini. The radio emission from the latter objects is consistent with that expected from an expanding shell of ionized gas that fades away as the gas becomes attenuated. The central star of the Crab Nebula has been detected as a radio (and optical) pulsar.

Measurements from rockets, balloons, and spacecraft have revealed distinct X-ray sources outside the solar system. The strongest galactic source, Scorpius X-1, appears to be associated with a hot variable star resembling an old nova. In all likelihood this is a binary star system containing a low-mass normal star and a nonluminous companion.

A number of globular clusters are sources of cosmic X-rays. Some of this X-ray emission appears as intense fluctuations of radiation lasting only a few seconds but changing in strength by as much as 25 times. These X-ray sources have become known as bursters, and several such objects have been discovered outside of globular clusters as well. Some bursters vary on a regular basis, while others seem to turn on and off randomly. The most popular interpretation holds that bursters are the result of binary systems in which one of the objects—a compact neutron star or black hole—pulls matter from the companion, a normal star. This matter is violently heated in the process, giving rise to X-rays. That the emission is often in the form of a burst is probably caused by something interrupting the flow of matter onto (or into) the compact object or by an eclipsing orbit of the binary system.

Stellar Structure

Stellar Atmospheres

To interpret a stellar spectrum quantitatively, knowledge of the variation in temperature and density with depth in the star's atmosphere is needed.

The gradient of temperature in a star's atmosphere depends on the method of energy transport to the surface. One way to move energy from the interior of a star to its surface is via radiation; photons produced in the core are repeatedly absorbed and reemitted by stellar atoms, gradually propagating to the surface. A second way is via convection, which is a nonradiative mechanism involving a physical upwelling of matter much as in a pot of boiling water. For the Sun, at least, there are ways of distinguishing the mechanism of energy transport.

Photographs of the Sun's disk show that the centre of the disk is brighter than the limb. The difference in brightness depends on the wavelength of the radiation detected: it is large in violet light, is small in red light, and nearly vanishes when the Sun is imaged in infrared radiation. This limb darkening arises because the Sun becomes hotter toward its core. At the centre of the disk, radiation is received from deeper and hotter layers (on average) instead of from the limb, and the dependence of temperature on depth can be shown to correspond to the transport of energy by radiation, not by convection, at least in the outer layers of the Sun's atmosphere.

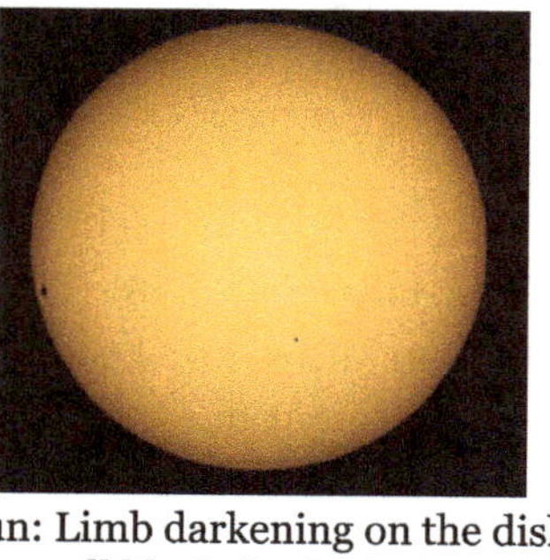

Limb darkening in the Sun: Limb darkening on the disk of the Sun. The planet Mercury can be seen as a small black dot in the lower middle of the solar disk.

The amount of limb darkening in any star depends on the effective temperature of the star and on the variation in temperature with depth. Limb darkening is occasionally an important factor in the analysis of stellar observations. For example, it must be taken into account to interpret properly the observed light curves of eclipsing binaries, and here again the results suggest transport of energy via radiation.

The layers of a normal star are assumed to be in mechanical, or hydrostatic, equilibrium. This means that at each point in the atmosphere, the pressure supports the weight of the overlying layers. In this way, a relation between pressure and density can be found for any given depth.

In addition to the temperature and density gradients, the chemical composition of the atmospheric layers as well as the opacity of the material must be known. In the Sun the principal source of opacity is the negative hydrogen ion (H^-), a hydrogen atom with one extra electron loosely bound to it. In the atmospheres of many stars, the extra electrons break loose and recombine with other ions, thereby causing a reemission of energy in the form of light. At visible wavelengths the main contribution to the opacity comes from the destruction of this ion by interaction with a photon (the above-cited process is termed photodissociation). In hotter stars, such as Sirius A (the temperature of which is about 10,000 K), atomic hydrogen is the main source of opacity, whereas in cooler stars much of the outgoing energy is often absorbed by molecular bands of titanium oxide, water vapour, and carbon monoxide. Additional sources of opacity are absorption by helium atoms and electron scattering in hotter stars, absorption by hydrogen molecules and molecular ions, absorption by certain abundant metals such as magnesium, and Rayleigh scattering (a type of wavelength-dependent scattering of radiation by particles named for the British physicist Lord Rayleigh) in cool supergiant stars.

At considerable depths in the Sun and similar stars, convection sets in. Though most models of stellar atmospheres (particularly the outer layers) assume plane-parallel stratified layers, photographs of granulation on the Sun's visible surface belie this simple picture. Realistic models must allow for rising columns of heated gases in some areas and descent of cooler gases in others. The motions of the radiating gases are especially important when the model is to be used to calculate the anticipated line spectrum of the star. Typical gas velocities are on the order of 2 km (1.4 miles) per second in the Sun; in other stars they can be much larger.

Temperature, density, and pressure all increase steadily inward in the Sun's atmosphere. The Sun has no distinct solid surface, so the point from which the depth or height is measured is arbitrary. The temperature of the visible layers ranges from 4,700 to 6,200 K, the density from about 10^{-7} to 4×10^{-7} gram per cubic cm, and the gas pressure from 0.002 to 0.14 atmospheres. The visible layers of stars such as the Sun have very low densities and pressures compared with Earth's atmosphere, even though the temperature is much higher. The strata of the solar atmosphere are very opaque compared with the terrestrial atmosphere.

For stars other than the Sun, the dependence of temperature on depth cannot be directly determined. Calculations must proceed by a process of successive approximations, during which the flux of energy is taken to be constant with depth. Computations have been undertaken for atmospheres of a variety of stars ranging from dwarfs to supergiants, from cool to hot stars. Their validity can be evaluated only by examining how well they predict the observed features of a star's continuous and line spectrum, including the detailed shapes of spectral-line features. Considering the known complexities of stellar atmospheres, the results fit the observations remarkably well.

Severe deviations exist for stars with extended and expanding atmospheres. Matter flowing outward from a star produces a stellar wind analogous to the solar wind, but one that is often much more extensive and violent. In the spectrum of certain very hot O-type stars (e.g., Zeta Puppis), strong, relatively narrow emission lines can be seen; however, in the ultraviolet, observations from rockets and spacecraft show strong emission lines with distinct absorption components on the shorter wavelength side. These absorption features are produced by rapidly outflowing atoms that absorb the radiation from the underlying stellar surface. The observed shifts in frequency correspond to ejection velocities of about 100 km (60 miles) per second. Much gentler stellar winds are found in cool M-type supergiants.

Rapid stellar rotation also can modify the structure of a star's atmosphere. Since effective gravity is much reduced near the equator, the appropriate description of the atmosphere varies with latitude. Should the star be spinning at speeds near the breakup point, rings or shells may be shed from the equator.

Some of the most extreme and interesting cases of rotational effects are found in close binary systems. Interpretations of the light and velocity curves of these objects suggest that the spectroscopic observations cannot be reconciled with simple orderly rotating stars. Instead, emission and absorption lines sometimes overlap in such a way as to suggest streams of gas moving between the stars. For example, Beta Lyrae, an eclipsing binary system, has a period of 12.9 days and displays very large shifts in orbital velocity. The brighter member at visible wavelengths is a star of type B6–B8; the other member is a larger, early B-type star that is embedded in an accretion disk and is draining matter from the B6–B8 star. The spectrum of the B6–B8-type component shows the regular velocity changes expected of a binary star, but there is an absorption (and associated emission) spectrum corresponding to a higher temperature (near spectral type B5) and a blue continuum corresponding to a very high-temperature star. The anomalous B5-type spectrum is from the accretion disk and is evidently excited principally by the star within it. This spectrum shows few changes in velocity with time.

Supergiant stars have very extended atmospheres that are probably not even approximately in hydrostatic equilibrium. The atmospheres of M-type supergiant stars appear to be slowly expanding outward. Observations of the eclipsing binary 31 Cygni show that the K-type supergiant component has an extremely inhomogeneous, extended atmosphere composed of numerous blobs and filaments. As the secondary member of this system slowly moves behind the larger star, its light shines through larger masses of the K-type star's atmosphere. If the atmosphere were in orderly layers, the lines of ionized calcium, for example, produced by absorption of the light of the B-type star by the K-type star's atmosphere, would grow stronger uniformly as the eclipse proceeds. They do not, however.

Stellar Interiors

Models of the internal structure of stars—particularly their temperature, density, and pressure gradients below the surface—depend on basic principles. It is especially important that model calculations take account of the change in the star's structure with time as its hydrogen supply is gradually converted into helium. Fortunately, given that most stars can be said to be examples of an "ideal gas," the relations between temperature, density, and pressure have a basic simplicity.

Distribution of Matter

Several mathematical relations can be derived from basic physical laws, assuming that the gas is "ideal" and that a star has spherical symmetry; both these assumptions are met with a high degree of validity. Another common assumption is that the interior of a star is in hydrostatic equilibrium. This balance is often expressed as a simple relation between pressure gradient and density. A second relation expresses the continuity of mass; i.e., if M is the mass of matter within a sphere of radius r, the mass added, ΔM, when encountering an increase in distance Δr through a shell of volume $4\pi r^2\Delta r$, equals the volume of the shell multiplied by the density, ρ. In symbols, $\Delta M = 4\pi r^2 \rho \Delta r$.

A third relation, termed the equation of state, expresses an explicit relation between the temperature, density, and pressure of a star's internal matter. Throughout the star the matter is entirely gaseous, and, except in certain highly evolved objects, it obeys closely the perfect gas law. In such neutral gases the molecular weight is 2 for molecular hydrogen, 4 for helium, 56 for iron, and so on. In the interior of a typical star, however, the high temperatures and densities virtually guarantee that nearly all the matter is completely ionized; the gas is said to be plasma, the fourth state of matter. Under these conditions not only are the hydrogen molecules dissociated into individual atoms, but also the atoms themselves are broken apart (ionized) into their constituent protons and electrons. Hence, the molecular weight of ionized hydrogen is the average mass of a proton and an electron—namely, 1/2 on the atom-mass scale noted above. By contrast, a completely ionized helium atom contributes a mass of 4 with a helium nucleus (alpha particle) plus two electrons of negligible mass; hence, its average molecular weight is 4/3. As another example, a totally ionized nickel atom contributes a nucleus of mass 58.7 plus 28 electrons; its molecular weight is then 58.7/29 = 2.02. Since stars contain a preponderance of hydrogen and helium that are completely ionized throughout the interior, the average particle mass, μ, is the (unit) mass of a proton, divided by a factor taking into account the concentrations by weight of hydrogen, helium, and heavier ions. Accordingly, the molecular weight depends critically on the star's chemical composition, particularly on the ratio of helium to hydrogen as well as on the total content of heavier matter.

If the temperature is sufficiently high, the radiation pressure, P_r, must be taken into account in addition to the perfect gas pressure, P_g. The total equation of state then becomes $P = P_g + P_r$. Here P_g depends on temperature, density, and molecular weight, whereas P_r depends on temperature and on the radiation density constant, $a = 7.5 \times 10^{-15}$ ergs per cubic cm per degree to the fourth power. With $\mu = 2$ (as an upper limit) and $\rho = 1.4$ grams per cubic cm (the mean density of the Sun), the temperature at which the radiation pressure would equal the gas pressure can be calculated. The answer is 28 million K, much hotter than the core of the Sun. Consequently, radiation pressure may be neglected for the Sun, but it cannot be ignored for hotter, more massive stars. Radiation pressure may then set an upper limit to stellar luminosity.

Certain stars, notably white dwarfs, do not obey the perfect gas law. Instead, the pressure is almost entirely contributed by the electrons, which are said to be particulate members of a degenerate gas. If μ' is the average mass per free electron of the totally ionized gas, the pressure, P, and density, ρ, are such that P is proportional to a 5/3 power of the density divided by the average mass per free electron; i.e., $P = 10^{13}(\rho/\mu')^{5/3}$. The temperature does not enter at all. At still higher densities the equation of state becomes more intricate, but it can be shown that even this complicated equation of state is adequate to calculate the internal structure of the white dwarf stars. As a result, white dwarfs are probably better understood than most other celestial objects.

For normal stars such as the Sun, the energy-transport method for the interior must be known. Except in white dwarfs or in the dense cores of evolved stars, thermal conduction is unimportant because the heat conductivity is very low. One significant mode of transport is an actual flow of radiation outward through the star. Starting as gamma rays near the core, the radiation is gradually "softened" (becomes longer in wavelength) as it works its way to the surface (typically, in the Sun, over the course of about a million years) to emerge as ordinary light and heat. The rate of flow of radiation is proportional to the thermal gradient—namely, the rate of change of temperature with interior distance. Providing yet another relation of stellar structure, this equation uses the following important quantities: a, the radiation constant noted above; c, the velocity of light; ρ, the density; and κ, a measure of the opacity of the matter. The larger the value of κ, the lower the transparency of the material and the steeper the temperature fall required to push the energy outward at the required rate. The opacity, κ, can be calculated for any temperature, density, and chemical composition and is found to depend in a complex manner largely on the two former quantities.

In the Sun's outermost (though still interior) layers and especially in certain giant stars, energy transport takes place by quite another mechanism: large-scale mass motions of gases—namely, convection. Huge volumes of gas deep within the star become heated, rise to higher layers, and mix with their surroundings, thus releasing great quantities of energy. The extraordinarily complex flow patterns cannot be followed in detail, but when convection occurs, a relatively simple mathematical relation connects density and pressure. Wherever convection does occur, it moves energy much more efficiently than radiative transport.

Source of Stellar Energy

The most basic property of stars is that their radiant energy must derive from internal sources. Given the great length of time that stars endure (some 10 billion years in the case of the Sun), it can be shown that neither chemical nor gravitational effects could possibly yield the required energies. Instead, the cause must be nuclear events wherein lighter nuclei are fused to create heavier nuclei, an inevitable by-product being energy.

In the interior of a star, the particles move rapidly in every direction because of the high temperatures present. Every so often a proton moves close enough to a nucleus to be captured, and a nuclear reaction takes place. Only protons of extremely high energy (many times the average energy in a star such as the Sun) are capable of producing nuclear events of this kind. A minimum temperature required for fusion is roughly 10 million K. Since the energies of protons are proportional to temperature, the rate of energy production rises steeply as temperature increases.

For the Sun and other normal main-sequence stars, the source of energy lies in the conversion of hydrogen to helium. The nuclear reaction thought to occur in the Sun is called the proton-proton cycle. In this fusion reaction, two protons (^1H) collide to form a deuteron (a nucleus of deuterium, ^2H), with the liberation of a positron (the electron's positively charged antimatter counterpart, denoted e^+). Also emitted is a neutral particle of very small mass called a neutrino, ν. While the helium "ash" remains in the core where it was produced, the neutrino escapes from the solar interior within seconds. The positron encounters an ordinary negatively charged electron, and the two annihilate each other, with much energy being released. This annihilation energy amounts to 1.02 megaelectron volts (MeV), which accords well with Einstein's equation $E = mc^2$ (where m is the mass of the two particles, c the velocity of light, and E the liberated energy).

Next, a proton collides with the deuteron to form the nucleus of a light helium atom of atomic weight 3, ^3He. A "hard" X-ray (one of higher energy) or gamma-ray (γ) photon also is emitted. The most likely event to follow in the chain is a collision of this ^3He nucleus with a normal ^4He nucleus to form the nucleus of a beryllium atom of weight 7, ^7Be, with the emission of another gamma-ray photon. The ^7Be nucleus in turn captures a proton to form a boron nucleus of atomic weight 8, ^8B, with the liberation of yet another gamma ray.

The ^8B nucleus, however, is very unstable. It decays almost immediately into beryllium of atomic weight 8, ^8Be, with the emission of another positron and a neutrino. The nucleus itself thereafter decays into two helium nuclei, ^4He. These nuclear events can be represented by the following equations:

$$^3He + {}^4He \rightarrow {}^7Be + \gamma \quad \text{(rather slow reaction)}$$

$$^3Be + {}^1H \rightarrow {}^8B + \gamma \quad \text{(rapid reaction)}$$

$$\left.\begin{aligned} {}^8B &\rightarrow {}^8Be + e^+ + \upsilon \\ {}^8Be &\rightarrow 2\,{}^4He \end{aligned}\right\} \quad \text{(instantaneous reactions)}$$

In the course of these reactions, four protons are consumed to form one helium nucleus, while two electrons perish.

The mass of four hydrogen atoms is 4 × 1.00797, or 4.03188, atomic mass units; that of a helium atom is 4.0026. Hence, 0.02928 atomic mass unit, or 0.7 percent of the original mass, has disappeared. Some of this has been carried away by the elusive neutrinos, but most of it has been converted to radiant energy. In order to keep shining at its present rate, a typical star (e.g., the Sun) needs to convert 674 million tons of hydrogen to 670 million tons of helium every second. According to the formula $E = mc^2$, more than four million tons of matter literally disappear into radiation each second.

This theory provides a good understanding of solar energy generation, although for decades it suffered from one potential problem. The neutrino flux from the Sun was measured by different experimenters, and only one-third of the flux of electron neutrinos predicted by the theory was detected. (Neutrinos come in three types, each associated with a charged lepton: electron neutrino, muon neutrino, and tau neutrino.) Over that time, however, the consensus grew that the solar neutrino problem and its solution lay not with the astrophysical model of the Sun but with the physical nature of neutrinos themselves. In the late 1990s and the early 21st century, scientists discovered that neutrinos oscillate (change types) between electron neutrinos, the state in which they were created in the Sun, and muon or tau neutrinos, states that are more difficult to detect when they reach Earth.

The main source of energy in hotter stars is the carbon cycle (also called the CNO cycle for carbon, nitrogen, and oxygen), in which hydrogen is transformed into helium, with carbon serving as a catalyst. The reactions proceed as follows: first, a carbon nucleus, ^{12}C, captures a proton (hydrogen nucleus), ^{1}H, to form a nucleus of nitrogen, ^{13}N, a gamma-ray photon being emitted in the process; thus, $^{12}C + {}^{1}H \rightarrow {}^{13}N + \gamma$. The light ^{13}N nucleus is unstable, however. It emits a positron, e^+, which encounters an ordinary electron, e^-, and the two annihilate one another. A neutrino also is released, and the resulting ^{13}C nucleus is stable. Eventually the ^{13}C nucleus captures another proton, forms ^{14}N, and emits another gamma-ray photon. In symbols the reaction is represented by the equations,

$$^{13}N \rightarrow {}^{13}C + e^+ + \nu;$$

$$\text{then } {}^{13}C + {}^{1}H \rightarrow {}^{14}N + \gamma.$$

Ordinary nitrogen, ^{14}N, is stable, but when it captures a proton to form a nucleus of light oxygen-15, ^{15}O, the resulting nucleus is unstable against beta decay. It therefore emits a positron and a neutrino, a sequence of events expressed by the equations,

$$^{14}N + {}^{1}H \rightarrow {}^{15}O + \gamma;$$

$$\text{then } {}^{15}O \rightarrow {}^{15}N + e^+ + \nu.$$

Again, the positron meets an electron, and the two annihilate each other while the neutrino escapes. Eventually the ^{15}N nucleus encounters a fast-moving proton, ^{1}H, and captures it, but the formation of an ordinary ^{16}O nucleus by this process occurs only rarely. The most likely effect of this proton capture is a breakdown of ^{15}N and a return to the ^{12}C nucleus—that is,

$$^{15}N + {}^{1}H \rightarrow {}^{12}C + {}^{4}He + \gamma.$$

Thus, the original ^{12}C nucleus reappears, and the four protons that have been added permit the formation of a helium nucleus. The same amount of mass has disappeared, though a different fraction of it may have been carried off by the neutrinos.

Only the hottest stars that lie on the main sequence shine with energy produced by the carbon cycle. The faint red dwarfs use the proton-proton cycle exclusively, whereas stars such as the Sun shine mostly by the proton-proton reaction but derive some contribution from the carbon cycle as well.

The aforementioned mathematical relationships permit the problem of stellar structure to be addressed notwithstanding the complexity of the problem. An early assumption that stars have a uniform chemical composition throughout their interiors simplified the calculations considerably, but it had to be abandoned when studies in stellar evolution proved that the compositions of stars change with age. Computations need to be carried out by a step-by-step process known as numerical integration. They must take into account that the density and pressure of a star vanish at the surface, whereas these quantities and the temperature remain finite at the core.

Resulting models of a star's interior, including the relation between mass, luminosity, and radius, are determined largely by the mode of energy transport. In the Sun and the fainter main-sequence stars, energy is transported throughout the outer layers by convective currents, whereas in the deep interior, energy is transported by radiation. Among the hotter stars of the main sequence, the reverse appears to be true. The deep interiors of the stars that derive their energy primarily from the carbon cycle are in convective equilibrium, whereas in the outer parts the energy is carried by radiation. The observed masses, luminosities, and radii of most main-sequence stars can be reproduced with reasonable and uniform chemical composition.

Chemically homogeneous models of giant and supergiant stars cannot be constructed. If a yellow giant such as Capella is assumed to be built like a main-sequence star, its central temperature turns out to be so low that no known nuclear process can possibly supply the observed energy output. Progress has been made only by assuming that these stars were once main-sequence objects that, in the course of their development, exhausted the hydrogen in their deep interiors. Inert cores consequently formed, composed mainly of the helium ash left from the hydrogen-fusion process. Since no helium nuclear reactions are known to occur at the few tens of millions of kelvins likely to prevail in these interiors, no thermonuclear energy could be released from such depleted cores. Instead, energy is assumed to be generated in a thin shell surrounding the inert core where some fuel remains, and it is presumably produced by the carbon cycle. Such models are called shell-source models. As a star uses up increasing amounts of its hydrogen supply, its core grows in mass while the outer envelope of the star continues to expand. These shell-source models explain the observed luminosities, masses, and radii of giants and supergiants.

The depletion of hydrogen fuel is appreciable even for a dwarf, middle-aged star such as the Sun. The Sun seems to have been shining at its present rate for about the last 20 percent of its current age of five billion years. For its observed luminosity to be maintained, the Sun's central temperature must have increased considerably since the formation of the solar system, largely as a consequence of the depletion of the hydrogen in its interior along with an accompanying increase in molecular weight and temperature. During the past five billion years, the Sun probably brightened by about half a magnitude; in early Precambrian time (about two billion years ago), the solar luminosity must have been some 20 percent less than it is today.

Star Formation

Stars are the fundamental units of luminous matter in the universe, and they are responsible, directly or indirectly, for most of what we see when we observe it. They also serve as our primary tracers of the structure and evolution of the universe and its contents. Consequently, it is of central importance in astrophysics to understand how stars form and what determines their properties. The generally accepted view that stars form by the gravitational condensation of diffuse matter in space is very old, indeed almost as old as the concept of universal gravitational attraction itself, having been suggested by Newton in 1692. However, it is only in the past half-century that the evidence has become convincing that stars are presently forming by the condensation of diffuse interstellar matter in our Galaxy and others, and it is only in recent decades that we have begun to gain some physical understanding of how this happens. Observations at many wavelengths, especially radio and infrared, have led to great advances in our knowledge of the subject, and the observational study of star formation is now a large and active field of research. Extensive theoretical and computational work has also contributed increasingly to clarifying the physical processes involved.

Star formation occurs as a result of the action of gravity on a wide range of scales, and different mechanisms may be important on different scales, depending on the forces opposing gravity. On galactic scales, the tendency of interstellar matter to condense under gravity into star-forming clouds is counteracted by galactic tidal forces, and star formation can occur only where the gas becomes dense enough for its self-gravity to overcome these tidal forces, for example in spiral arms. On the intermediate scales of star-forming 'giant molecular clouds' (GMCs), turbulence and magnetic fields may be the most important effects counteracting gravity, and star formation may involve the dissipation of turbulence and magnetic fields. On the small scales of individual prestellar cloud cores, thermal pressure becomes the most important force resisting gravity, and it sets a minimum mass that a cloud core must have to collapse under gravity to form stars. After such a cloud core has begun to collapse, the centrifugal force associated with its angular momentum eventually becomes important and may halt its contraction, leading to the formation of a binary or multiple systems of stars. When a very small central region attains stellar density, its collapse is permanently halted by the increase of thermal pressure and an embryonic star or 'protostar' forms and continues to grow in mass by accretion. Magnetic fields may play a role in this final stage of star formation, both in mediating gas accretion and in launching the bipolar jets that typically announce the birth of a new star.

In addition to these effects, interactions between the stars in a forming multiple system or cluster may play an important role in the star formation process. Most, and possibly all, stars form with close companions in binary or multiple systems or clusters, and gravitational interactions between the stars and gas in these systems may cause the redistribution of angular momentum that is necessary for stars to form. Interactions in dense environments, possibly including direct stellar collisions and mergers, may play a particularly important role in the formation of massive stars. Such processes, instead of generating characteristic properties for formation 1653 forming stars may be chaotic and create a large dispersion in the properties of stars and stellar systems. Thus, star formation processes, like most natural phenomena, probably involve a combination of regularity and randomness.

Some outcomes of star formation processes that are particularly important to understand include the rate at which the gas in galaxies is turned into stars, and the distribution of masses with which stars are formed. The structures of galaxies depend on the circumstances in which stars form and the rate at which they form, while the evolution of galaxies depends on the spectrum of masses with which they form, since low-mass stars are faint and evolve slowly while massive ones evolve fast and release large amounts of matter and energy that can heat and ionize the interstellar gas, enrich it with heavy elements, and possibly expel some of it into intergalactic space. It is also important to understand the formation of binary systems because many important astrophysical processes, including the formation of various kinds of exotic objects, involve the interactions of stars in binary systems. A further outcome of star formation that is of great interest to understand is the formation of planetary systems, which may often form as byproducts of star formation in disks of leftover circumstellar material.

Observed Properties of Star-forming Clouds

Sites of Star Formation

Most of the star formation in galaxies occurs in spiral arms, which are marked primarily by their concentrations of luminous young stars and associated glowing clouds of ionized gas. The most luminous stars have lifetimes shorter than 10 Myr, or 10^{-3} times the age of the universe, so they must have formed very recently from the dense interstellar gas that is also concentrated in the spiral arms. Star formation occurs also near the centres of some galaxies, including our own Milky Way galaxy, but this nuclear star formation is often obscured by interstellar dust and its existence is inferred only from the infrared radiation emitted by dust heated by the embedded young stars. The gas from which stars form, whether in spiral arms or in galactic nuclei, is concentrated in massive and dense 'molecular clouds' whose hydrogen is nearly all in molecular form. Some nearby molecular clouds are seen as 'dark clouds' against the bright background of the Milky Way because their interstellar dust absorbs the starlight from the more distant stars.

In some nearby dark clouds many faint young stars are seen, most distinctive among which are the T Tauri stars, whose variability, close association with the dark clouds, and relatively high luminosities for their temperatures indicate that they are extremely young and have ages of typically only about 1 Myr. These T Tauri stars are the youngest known visible stars, and they are 'pre-main-sequence' stars that have not yet become 1654 R B Larson hot enough at their centres to burn hydrogen and begin the main-sequence phase of evolution. Some of these young stars are embedded in particularly dense small dark clouds, which are thus the most clearly identified sites of star formation. These clouds have been studied extensively, using radio techniques to observe the heavier molecules such as CO and infrared techniques to study the dust. Observations of the thermal emission from the dust at far-infrared wavelengths have proven to be particularly useful for studying the structure of these small star-forming clouds; the dust is the best readily observable tracer of the mass distribution because most of the heavier molecules freeze out onto the dust grains at high densities. Many of these small clouds are dense enough for gravity to hold them together against pressure and cause them to collapse into stars, strengthening their identification as stellar birth sites.

The smaller and more isolated dark clouds have received the most attention because they are easiest to study and model, but most stars actually form in larger groups and clusters and in larger and more complex concentrations of molecular gas. There is no clear demarcation between molecular concentrations of different size, and no generally accepted terminology for them, but the terms 'cloud', 'clump', and 'core' have all often been used, generally with reference to structures of decreasing size. The term 'clump' will be used to denote any region of enhanced density in a larger cloud, while the term 'core' will be used to denote a particularly dense self-gravitating clump that might collapse to form a star or group of stars. The term 'globule' has also been used to describe some compact and isolated dark clouds whose importance as stellar birth sites was advocated by Bok before their role in star formation was established; Bok's enthusiastic advocacy of these globules as sites of star formation has since been vindicated, and we now know that some of them are indeed forming stars.

Structure of Molecular Clouds

Surveys of the molecular gas in galaxies show that it is typically concentrated in large complexes or spiral arm segments that have sizes up to a kiloparsec and masses up to 10^7 solar masses ($M_\odot$). These complexes may contain several GMCs with sizes up to 100 parsecs (pc) and masses up to 10^6 M, and these GMCs, in turn, contain much smaller scale structure that may be filamentary or clumpy on a wide range of scales. The substructures found in GMCs range from massive clumps with sizes of a several parsecs and masses of thousands of solar masses, which may form entire clusters of stars, to small dense cloud cores with sizes of the order of 0.1 pc and masses of the order of 1 $M_\odot$, which may form individual stars or small multiple systems. The internal structure of molecular clouds is partly hierarchical, consisting of smaller subunits within larger ones, and fractal models may approximate some aspects of this structure. In particular, the irregular boundaries of molecular clouds have fractal-like shapes resembling those of surfaces in turbulent flows, and this suggests that the shapes of molecular clouds may be created by turbulence.

Molecular clouds are the densest parts of the interstellar medium, and they are surrounded by less dense envelopes of atomic gas. The abundance of molecules increases with density because hydrogen molecules form on the surfaces of dust grains and the rate of this process increases with increasing density. In addition, the survival of the molecules requires that the clouds have a sufficient opacity due to dust to shield the molecules from ultraviolet radiation capable of dissociating them, and this means that molecular clouds must have a column density of at least 20 M pc^{-2}. Most molecular clouds have column densities much higher than this and are therefore quite opaque, a typical column density being of the order of 100 $M_\odot$ pc^{-2}. Because of the high densities of molecular clouds, the rate at which they are cooled by collisionally excited atomic and molecular emission processes is high, and because of their high opacity, the rate at which they are heated by external radiation is low; the result is that molecular clouds are very cold and have typical temperatures of only about 10–20 K. Higher temperatures of up to 100 K or more may exist locally in regions heated by luminous newly formed stars. In typical molecular clouds, cooling is due mostly to the emission of far-infrared radiation from molecules such as CO, which is usually the most important coolant. However, in the densest collapsing cloud cores the gas becomes thermally coupled to the dust, which then controls the temperature by its strongly temperature-dependent thermal emission, maintaining a low and almost constant temperature of about 10 K over a wide range of densities. This low and nearly constant temperature is an important feature of the star formation process, and is what makes possible the collapse of prestellar cloud cores with masses as small as one solar mass or less.

The gas densities in molecular clouds vary over many orders of magnitude: the average density of an entire GMC may be of the order of 20 H_2 molecules per cm^3, while the larger clumps within it may have average densities of the order of 10^3 H_2 cm^{-3} and the small prestellar cloud cores may have densities of 10^5 H_2 cm^{-3} or more. At the high densities and low temperatures characteristic of molecular clouds, self-gravity is important and it dominates over thermal pressure by a large factor, except in the smallest clumps. If thermal pressure were the only force opposing gravity, molecular clouds might then be expected to collapse rapidly and efficiently into stars. Most molecular clouds are indeed observed to be forming stars, but they do so only very inefficiently, typically turning only a few percent of their mass into stars before being dispersed. The fact that molecular clouds do not quickly turn most of their mass into stars, despite the strong dominance of gravity over thermal pressure, has long been considered problematic, and has led to the widely held view that additional effects such as magnetic fields or turbulence support these clouds in near-equilibrium against gravity and prevent a rapid collapse.

However, the observed structure of molecular clouds does not resemble any kind of equilibrium configuration, but instead is highly irregular and filamentary and often even windblown in appearance, suggesting that these clouds are actually dynamic and rapidly changing structures, just like terrestrial clouds. The complex structure of molecular clouds is important to understand because it may influence or determine many of the properties with which stars and systems of stars are formed. For example, stars often appear to form in a hierarchical arrangement consisting of smaller groupings within larger ones, and this may reflect the hierarchical and perhaps fractal-like structure of star-forming clouds. Stars may also derive their masses directly from those of the prestellar cloud cores, as is suggested by the fact that the distribution of masses or 'initial mass function' (IMF) with which stars are formed appears to resemble the distribution of masses of the prestellar cores in molecular clouds.

The Role of Turbulence and Magnetic Fields

In addition to their irregular shapes, molecular clouds have complex internal motions, as is indicated by the broad and often complex profiles of their molecular emission lines. In all but the smallest clumps, these motions are supersonic, with velocities that significantly exceed the sound speed of 0.2 km s^{-1} typical for dark clouds.

The broad line profiles appear to reflect mostly small-scale random motions rather than largescale systematic motions such as cloud rotation, since observed large-scale motions are usually too small to contribute much to the line widths. The internal random motions in molecular clouds are often referred to as 'turbulence', even though their detailed nature remains unclear and they may not closely resemble classical eddy turbulence. Nevertheless, the existence of some kind of hierarchy of turbulent motions is suggested by the fact that the velocity dispersion inferred from the linewidth increases systematically with region size in a way that resembles the classical Kolmogoroff law. Supersonic turbulence may play an important role in structuring molecular clouds, since supersonic motions can generate shocks that produce large density fluctuations. Much effort has, therefore, been devoted to studying the internal turbulent motions in molecular clouds, and 'size–linewidth relations' have been found in many studies, albeit with considerable variability and scatter.

Studies that include motions on larger scales suggest that a similar correlation between velocity dispersion and region size extends up to galactic scales, and this suggests that the turbulent motions in molecular clouds are part of a larger-scale hierarchy of interstellar turbulent motions. Molecular clouds must then represent condensations in a generally turbulent interstellar medium, and their structure and dynamics must constitute part of the structure and dynamics of the medium as a whole. The origin of the observed large-scale interstellar motions is not yet fully understood, but many different sources almost certainly contribute to them, and the sources and properties of the interstellar turbulence may vary from place to place. Some likely sources include gravitationally driven motions on large scales and stellar feedback effects such as ionization, winds, and supernova explosions on smaller scales. Different star-forming regions do indeed show different levels of turbulence; for example, there is a higher level of turbulence in the Orion region, where stars of all masses are forming vigorously in two large GMCs, than in the smaller and more quiescent Taurus clouds, which are forming only less massive stars. In many cases, self-gravity is roughly balanced by the turbulent motions in molecular clouds, and this suggests that gravity and turbulence are equally important in controlling the structure and evolution of these clouds.

In addition to being turbulent, molecular clouds are also significantly magnetized, and magnetic fields can also be important for the dynamics and evolution of these clouds. If molecular clouds are sufficiently strongly magnetized, their internal motions might be predominantly wavelike, consisting basically of magnetohydrodynamic (MHD) waves such as Alfvén waves; since Alfvén waves involve only ´transverse and non-compressional motions, they might be expected to dissipate more slowly than purely hydrodynamic supersonic turbulence. Wavelike 'MHD turbulence' might then provide a source of pressure that can supplement thermal pressure and help to support molecular clouds for a significant time against gravity. If molecular clouds can indeed be supported against gravity for a long time by a combination of static magnetic fields and slowly dissipating MHD turbulence, this might provide some justification for models that treat these clouds as long-lived quasi-equilibrium structures.

It is difficult to test models that assume important magnetic cloud support because of the paucity of accurate measurements of field strengths; direct measurements using the Zeeman Effect are difficult and in most cases have yielded only upper limits.

A compilation of results by Crutcher suggests that static magnetic fields are not sufficient by themselves to counterbalance gravity, but that there may be a rough equipartition between the magnetic energy and the turbulent kinetic energy in molecular clouds, in which case a combination of static magnetic fields and MHD turbulence might together be able to support these clouds against gravity. Bourke et al find, with further data, that the measured magnetic field is not sufficient to balance gravity if the clouds studied are spherical but could be sufficient if these clouds are flattened and the undetected fields are close to their upper limits. If static magnetic fields were to be important in supporting molecular clouds against gravity, however, one might expect to see alignments between cloud structures and the magnetic field direction inferred from polarization studies, but efforts to find such alignments have yielded ambiguous results and often do not show the expected alignments.

Progress in understanding the role of MHD turbulence in molecular clouds has in the meantime come from numerical simulations. These simulations show that, even if magnetic fields are important and the turbulence is predominantly wavelike, any MHD wave motions decay in a time comparable to the dynamical or crossing time of a cloud, defined as the cloud size divided by a typical turbulent velocity. This rapid wave dissipation occurs because, even if all of the wave energy is initially in transverse motions, motions along the field lines are immediately generated and they produce shocks that soon dissipate the wave energy. Thus, gravity cannot be balanced for long by any kind of turbulence unless the turbulence is continually regenerated by a suitable energy source. But fine tuning is then needed, and sufficient turbulence must be generated on all scales to maintain a cloud in equilibrium without disrupting it. Models of this kind are constrained by the fact that the heating associated with the dissipation of internally generated turbulence may produce temperatures higher than are observed.

If molecular clouds are in fact transient rather than quasi-equilibrium structures, as is suggested by the evidence on their lifetimes, this removes much of the motivation for postulating long-term magnetic support. Magnetic fields may still have important consequences for star formation, but they may exert their most important effects during the early phases of cloud evolution when the fields are still strongly coupled to the gas and can damp rotational motions, thus helping to solve the 'angular momentum problem' of star formation. During the later high-density stages of collapse, the gas is expected to decouple from the magnetic field by ambipolar diffusion, and the magnetic field then becomes dynamically unimportant until a much later stage when stellar conditions are approached at the centre.

Cloud Evolution and Lifetimes

Evidence concerning the lifetimes and evolution of molecular clouds is provided by the ages of the associated newly formed stars and star clusters. Very few GMCs are known that are not forming stars, and the most massive and dense ones all contain newly formed stars. This means that there cannot be any significant 'dead time' between the formation of a massive dense molecular cloud and the onset of star formation in it; in particular, there cannot be any long period of slow quasi-static evolution before stars begin to form. Molecular clouds also cannot survive for long after beginning to make stars, since the age span of the associated young stars and clusters is never more than about 10 Myr, about the dynamical or crossing time of a large GMC, and since stars and clusters older than 10 Myr no longer have any associated 1658 R B Larson molecular gas. Smaller clouds have even shorter stellar age spans that in all cases are comparable to their crossing times; in the Taurus clouds, for example, most of the stars have formed just in the past few million years. On the smallest scales, prestellar cloud cores have yet shorter estimated lifetimes that are only a few hundred thousand years for the smallest and densest cores. Star formation is, therefore, evidently a fast process, and it always occurs in a time comparable to the crossing time of the associated molecular cloud or core. After that, a star-forming cloud must soon be destroyed or become no longer recognizable, perhaps being destroyed by stellar feedback effects such as ionization or being dispersed or restructured by larger-scale interstellar flows.

Studies of the chemistry of molecular clouds also suggest young ages and short lifetimes. As discussed by Stahler, Prasad et al, Herbst, and van Dishoeck and Blake, molecular clouds and cloud cores appear to be chemically relatively unevolved, since the observed abundances of various molecules are often far from those expected to prevail in chemical equilibrium and resemble instead those predicted to occur at an early stage of cloud evolution less than 1 Myr after their formation. This again suggests that molecular clouds are quite young, or at least that they have undergone recent chemical reprocessing by some major restructuring event; in the latter case the observed structures must still be of recent origin. The fact that most of the molecules in dense molecular clouds have not yet frozen out on the dust grains, as would be expected if these clouds are stable long-lived objects, also suggests that molecular clouds or cores may be relatively young, of the order of 1 Myr or less.

The evidence therefore suggests that molecular clouds are transient structures that form, evolve, and disperse rapidly, all in a time comparable to the crossing time of their internal turbulent motions. Numerical simulations of turbulence in molecular clouds also do not support the possibility that these clouds can be supported against gravity for a long time in a near-equilibrium state and suggest that the observed structures are quite transient. This is consistent with the irregular and often windblown appearances of molecular clouds.

If molecular clouds are indeed transient structures, they must be assembled rapidly by largerscale motions in the interstellar medium. The processes that may be involved in the formation and destruction of molecular clouds have been reviewed by Larson and Elmegreen, and the processes that may be responsible for the rapid formation of stars in them during their brief existence.

Formation of Binary Systems

Angular Momentum Problem

As has long been recognized, the amount of angular momentum in a typical star-forming cloud core is several orders of magnitude too large to be contained in a single star, even when rotating at breakup speed; this is the classical 'angular momentum problem' of star formation. The angular momentum problem remains despite the fact that star-forming cloud cores rotate more slowly than would be expected if they had formed from low-density gas with conservation of angular momentum; this slow rotation can plausibly be explained as a consequence of magnetic braking acting during the early low-density phases of cloud evolution when magnetic fields are still strongly coupled to the gas. Magnetic fields are, however, predicted to decouple from the gas by ambipolar diffusion long before stellar densities are reached, and angular momentum is then expected to be approximately conserved during the later stages of collapse. Observations of collapsing cloud cores confirm that angular momentum is approximately conserved on scales smaller than a few hundredths of a parsec. Therefore an important part of the angular momentum problem remains unsolved, since the typical observed angular momentum of prestellar cloud cores is still three orders of magnitude larger than the maximum amount that can be contained in a single star.

In principle, there are two ways to dispose of this excess angular momentum: (1) it could be transported to outlying diffuse material, for example by viscous transport processes in a protostellar accretion disk, thus allowing most of the mass to be accreted by a central star, or (2) much of the initial angular momentum could go into the orbital motions of the stars in a binary or multiple system. Transport processes in disks do not clearly offer a solution to the angular momentum problem because no adequate transport mechanism has yet been identified, and because even if such a mechanism could be identified, a residual disk would have to expand to a very large radius to absorb all of the angular momentum, and this would make the accretion time longer than the inferred ages of young stars. The alternative possibility, which has also long been recognized, is that the collapse of a prestellar cloud core typically produces a binary or multiple system and that much of the initial angular momentum goes into the orbital motions of the stars in such a system. Observations show that the great majority of stars do indeed form in binary or multiple systems; since the angular momentum of a typical cloud core is comparable to that of a wide binary it is plausible that at least the wider binaries can be formed directly by the fragmentation of rotating cloud cores. Even the minority of stars that are observed to be single can be accounted for if stars typically form in unstable triple systems that decay into a binary and a single star; this simple scenario would be consistent with the observed fact that about two-thirds of all stars are in binary systems while about one-third are single.

Formation of Binary and Multiple Systems

Much theoretical and numerical work has suggested that the formation of binary or multiple systems is the usual result of collapse with rotation, while the formation of single stars occurs only in special cases. In many cases, a cloud core with significant rotation may fragment during the isothermal phase of collapse and directly form a binary or multiple systems. A systematic study of this problem concludes that the occurrence of fragmentation during isothermal collapse depends primarily on the initial ratio of thermal to gravitational energy, which is related to the number of Jeans masses present, and only secondarily on the rotation rate; fragmentation is predicted if this ratio is less than about 0.5. If collapse continues into an adiabatic phase with an opaque core and a flattened rotating configuration forms, such as a disk or bar, most of the initial mass will then settle into this flattened configuration and it will become unstable to fragmentation unless there is efficient outward transport of angular momentum. An extensive exploration of this problem shows that, in a wide range of cases, the final outcome is the formation of a binary or multiple system, which can occur via the formation and breakup of a ring or bar configuration as well as by the fragmentation of a disk.

If a central star of significant mass forms with a residual disk of only modest mass, the final outcome is less clear, but recent simulations that include a detailed treatment of radiative cooling show that even in a disk of modest mass, fragmentation can occur and lead to the formation of a smaller companion object such as a massive planet or small star 1680 R B Larson. If collapse begins with initial configurations that are far from axisymmetric, for example if the initial configuration is filamentary, or if a collapsing cloud core is perturbed by interactions with other cloud cores in a forming group or cluster, the formation of a binary or multiple system is even more likely to occur. In the most detailed and realistic simulation of cloud collapse and fragmentation yet performed, which models the formation of a small cluster of stars in a turbulent and filamentary collapsing cloud most of the stars form in several subgroups, and within these groups numerous binary and multiple systems are formed; in fact, the dynamics of these groups is highly complex and chaotic, and binary and multiple systems are continually formed and disrupted. Circumstellar disks are also continually formed and disrupted, and although they appear rather ubiquitously, they may often have only a transient existence before being either fragmented or disrupted by interactions.

Observed binary systems show a large dispersion in their properties, and they have no strongly preferred values for parameters such as separation and eccentricity; this observed broad dispersion itself suggests a very dynamic and chaotic formation process. Although the wider binaries might plausibly be formed directly by the fragmentation of rotating cloud cores, the large spread in binary separations towards smaller values implies that additional processes must be involved in the formation of the closer binaries; these processes must have a stochastic element to account for the large dispersion in separations, and also a dissipative element to reduce the average energy and angular momentum of typical forming binaries. Purely stellar–dynamical interactions in groups and multiple systems can account for some of the observed spread in binary properties, but they cannot account for the closest observed systems. The detailed simulation of cloud collapse and fragmentation by Bate et al produces many close binary systems with separations smaller than would be predicted by simple arguments based on conservation of angular momentum, and Bate et al conclude that three distinct mechanisms are involved in the formation of these close binaries: (1) the continuing accretion of gas by a forming binary, which can shrink its orbit by a large factor if the accreted gas has a smaller specific angular momentum than the binary, as is usually the case; (2) the loss of angular momentum from a forming binary due to tidal interaction with circumbinary gas, for example with gas in a circumbinary disk; and (3) dynamical interactions in forming multiple systems that tend to extract angular momentum from the closer binary pairs. Clearly, many complex processes may be involved in the formation of binary and multiple systems and large-scale simulations are needed to address this problem in a systematic way and make statistical predictions that can be compared with observations.

Role of Tidal Interactions

If most stars form in binary or multiple systems, gravitational interactions between the orbiting protostars and residual gas will play an important role in redistributing angular momentum in the system. Tidal interactions, for example, can redistribute angular momentum in several mdifferent ways: (1) angular momentum can be extracted from a circumstellar disk by tidal interaction with a companion star in a binary or multiple system; (2) an object that forms in a circumstellar disk, such as a giant planet or a small companion star, can tidally extract angular momentum from the inner part of the disk and transfer it to the outer part; or (3) a forming binary system can lose orbital angular momentum by interaction with surrounding material, for example with matter in a circumbinary disk.

If each accreting protostar in a forming binary system has its own circumstellar disk, as might be expected and as occurs in numerical simulations, the tidal perturbing effect of the companion star on each disk produces a gravitational torque that extracts angular momentum from the disk and transfers it to the binary orbit. The transfer of angular momentum from the disk to the orbit can drive continuing accretion from each disk onto its central protostar, and this may be an important mechanism for driving accretion from disks onto forming stars. A tidal perturbation generates a two-armed spiral disturbance in a disk that propagates inward as an acoustic wave, and such a wave can transport angular momentum outward in the disk, possibly in conjunction with other mechanisms. In a binary system, tidal interactions will tend to be self-regulating because if too little angular momentum is removed from a circumstellar disk, the disk will expand towards the companion as it gains matter, increasing the strength of the interaction; conversely, if too much angular momentum is removed, the disk will shrink and the tidal effect will be weakened. Numerical simulations show that the circumstellar disks in a forming binary system settle into a quasi-steady state in which angular momentum is continually transferred from the disks to the binary orbit although the internal transport of angular momentum within the disks is not accurately modelled in these simulations and may result partly from an artificial viscosity. Tidal transport mechanisms may be of quite general importance in astrophysics, and may drive accretion flows in many types of binary systems containing disks.

If the orbit of a forming binary system is eccentric, as is true for most binaries, the tidal effect will be time dependent and will produce strong disturbances at each periastron passage, possibly causing episodes of rapid accretion onto one or both stars. For stars that form in multiple systems or clusters, protostellar interactions can be even more violent and chaotic and can cause circumstellar disks to be severely disturbed or even disrupted. In such situations, protostellar accretion rates will vary strongly with time, regardless of the detailed mechanisms involved, and protostars may gain much of their mass in discrete events or episodes. Some of the evidence suggests that protostellar accretion rates are indeed highly variable and that much of the accretion may occur in bursts. The observed luminosities of many protostars are lower than is predicted for models with steady accretion, and this can be understood if most of the accretion occurs in short bursts. The jet-like Herbig–Haro outflows, which are believed to be powered by rapid accretion onto forming stars at an early stage of evolution, are clearly episodic or pulsed, and this suggests that the accretion process is itself episodic. It is noteworthy that the jet sources are frequently found to have close stellar companions: at least 85% of the jet sources are members of binary or triple systems, which is the highest binary frequency yet found among young stellar objects. This fact strongly suggests that there is a causal connection between the presence of a close companion and the launching of a jet, as would be expected if tidal interactions were responsible for the episodes of rapid accretion that generate the jets.

The same bursts of rapid accretion that produce the jets might also be responsible for the 'FU Orionis' outbursts that are observed in some newly formed stars; these outbursts are also thought to be caused by episodes of rapid accretion. The possibility that the FU Orionis phenomenon is caused by tidal interactions in highly eccentric binary systems was first suggested by Toomre in 1985 (private communication, quoted by Kenyon et al and Hartmann and Kenyon and it is supported by the numerical simulations of Bonnell and Bastien which show bursts of rapid accretion triggered by tidal interactions in a forming binary system. However, the nature of the outburst itself remains to be clarified; the idea that it is due to the rapid brightening of a self-luminous accretion disk has been questioned by Herbig et al, who argue that there is no clear spectroscopic evidence that the observed radiation from FU Ori objects comes from disks, and suggest that these objects are better interpreted as rapidly rotating stars whose outer layers have been heated and expanded by some dynamical disturbance, as suggested by Larson. One source of such a disturbance could be an episode of rapid accretion from a disk, possibly induced by a tidal interaction with a companion star as suggested above; this is perhaps more plausible than the rotational instability originally suggested by Larson. If a causal connection could be established among protostellar interactions, FU Ori outbursts, and Herbig–Haro jets, this would provide strong evidence that tidal interactions play a central role in the star formation process.

Implications for Planet Formation

If protostellar interactions play an important role in star formation, this will clearly have important consequences for planet formation too. Quiescent disks like the 'solar nebula' in which our own planetary system is thought to have formed may in this case not be as common as has been assumed, and planet-forming disks may often be disturbed by interactions with companion stars in forming systems of stars. Even our own solar system may have experienced significant disturbances early in its history, as is suggested by the fact that the fundamental plane of our planetary system is tilted by 8° with respect to the Sun's equatorial plane; this tilt could plausibly have been caused by an encounter with another star in a young multiple system or cluster. If nearly all stars form in multiple systems, single stars with disks may mostly originate by ejection from such systems, as indeed happens in the detailed simulation of cluster formation by Bate et al. Our own planetary system might then represent only a somewhat accidental particular case and might not be typical, since chaotic star formation processes may produce a large dispersion in the properties of planet-forming disks.

Another consequence of a dynamic and chaotic picture of star formation in which circumstellar disks are often disturbed by interactions is that shocks generated by such disturbances can produce transient heating events that might account for the high-temperature inclusions observed in meteorites. Chondrules, for example, show evidence for recurrent short heating events that reach temperatures of the order of 2000 K. Shock waves in protostellar disks with speeds of several km s^{-1} could produce temperatures of the required order for the required short times, and thus could provide a viable heating mechanism. Shocks of this strength could plausibly be produced by tidal interactions in forming systems of stars, such as the frequent interactions seen in the simulation of cluster formation by Bate et al. The chondrules in meteorites might then bear evidence of a violent early history of our Solar system.

Formation of Massive Stars and Clusters

The formation of stars with larger masses clearly involves more complex processes and is less well understood, both observationally and theoretically. One difference between the formation of massive stars and that of low-mass stars is that the most massive stars have masses that are much larger than the Jeans mass in the cloud cores in which they form; although these more massive cores have higher temperatures than the smaller ones in which low-mass stars form, they also have much higher densities, and the result is that the Jeans mass is not very different from that in the low-mass cores. Thus, the large cloud cores in which massive stars form might be expected to contain many smaller bound clumps, as is indeed suggested by some of the evidence, so these cores might form groups or clusters of stars rather than individual stars or binary systems. Furthermore, the more massive cloud cores all contain supersonic internal turbulent motions which will generate large density fluctuations and clumpy substructure even if none was present initially, again suggesting that massive stars may form only in groups or clusters.

A second important difference is that the intense radiation emitted by a massive accreting protostar can produce feedback effects on the infalling envelope that limit or prevent continuing accretion, for example via the effects of radiative heating, radiation pressure, or ionization on the envelope. The most important of these feedback effects may be radiation pressure, and detailed calculations have found that the force exerted by radiation pressure in a spherical infalling envelope may exceed the force of gravity when the protostellar mass exceeds about 10 M. Models that assume spherical infall may, as a result, not be able to account for the formation of stars with masses much larger than 10 M, although they may suffice for stars up to this mass. If matter is instead accreted from a circumstellar disk, a protostar may be able to build up a larger mass before radiation pressure terminates the accretion; a detailed calculation by Yorke and Sonnhalter of collapse with rotation that incorporates radiation pressure and the formation of a viscous disk shows that in this type of model the mass of an accreting protostar may reach 30 M or more, although the star formation process is still very inefficient and most of the initial mass, and even most of the disk, are eventually dispersed by radiation pressure. Therefore, it appears that even more complex models may be needed to account for the formation of the most massive stars.

Because regions of massive star formation are more complex in structure and also more distant than the well-studied regions of low-mass star formation, it is more difficult to study their structure in detail and to identify individual star-forming units; these problems are compounded by the more complex chemistry of these regions and by their complex dynamics, which often includes multiple outflows. The densities and temperatures of massive star-forming cloud cores are both considerably higher than those of low-mass cores, and they may have densities exceeding 10^7 cm^{-3} and temperatures exceeding 100 K in some locations; however, it is not clear whether the gas with these observed properties is prestellar or has already been substantially compressed and heated by prior star formation activity. Therefore, the most useful evidence concerning how massive stars form may be the fossil record provided by the properties of newly formed massive stars and stellar systems. Young massive stars are almost always located in clusters or associations, which are often hierachical in structure and contain subgroups; the mass of the most massive star in each subgroup or cluster tends to increase systematically with the mass of the cluster. The most massive stars are often centrally located in the clusters in which they form, and this implies that these stars must have formed near the cluster centre, suggesting that accumulation processes have played an important role in their formation. The most massive stars also have an unusually high frequency of close companions in binary or multiple systems, and these companions are themselves often massive objects. The evidence thus clearly indicates that massive stars form in exceptionally dense and complex environments, typically at the centres of clusters and in close proximity to other stars that are themselves often massive.

Possible Formation Processes

Several authors have suggested that the formation of massive stars can be modelled with a suitable extension of the standard model for low-mass star formation that was developed by Shu et al. In support of this possibility, Garay and Lizano and Garay et al have noted that there is evidence suggesting that disks and outflows play a similar role in the formation of massive stars as they do in the formation of low-mass stars, although the existence of circumstellar disks is less well established for massive stars than for low-mass stars. These authors also note that if the ambient gas density and hence the accretion rate are several orders of magnitude higher for massive protostars than they are for low-mass protostars, as is suggested by the evidence, then the ram pressure of the inflow may overcome radiation pressure in the infalling envelope and allow continuing accretion to occur even for luminous objects.

On the other hand, the fact that massive stars tend to form near the centres of dense clusters with many close companions suggests that protostellar interactions may play a particularly important role in the formation of massive stars and may help to drive the required high mass accretion rates. An extreme type of interaction that can occur in a sufficiently dense environment is direct protostellar collisions and mergers, and such mergers have been suggested to play an important role in building up the most massive stars. Mergers would clearly be the most effective way of overcoming radiation pressure and other feedback effects, and they might also help to account for the high frequency of close companions of massive stars as the results of failed mergers. Even though extremely high densities are required, mergers must sometimes occur because some binary systems are observed that contain stars that are almost in contact; if nature can make stars that are almost in contact, it must surely make some that come even closer together and merge. Given the existence of the two apparently competing hypotheses, it has been debated whether massive stars form by 'accretion' or by 'mergers' but the implied dichotomy between accretion and mergers is almost certainly oversimplified, and both types of processes probably play a role in the formation of massive stars.

Important progress has come from simulations of the formation of stars with a wide range of masses in forming clusters of stars. In these simulations, most of the stars form in subgroups at the nodes of a filamentary network, a feature that appears to be a generic result of gravitational collapse on scales much larger than the Jeans length, Balsara et al, and these subgroups eventually merge into a single large cluster. A spectrum of protostellar masses is built up as protostars in different environments accrete mass at different rates, the more massive stars forming by faster runaway accretion processes in the denser environments and the larger subgroups. In the simulation by Bonnell et al of the formation of a cluster of about 400 objects in an isothermally collapsing cloud, the most massive object eventually acquires a mass of about 27 $M_{\odot}$ while in the simulation by Bonnell and Bate of a cluster of 1000 accreting protostars that gain mass by both accretion and mergers, the most massive object attains a mass of about 50 M, or 100 times the mass of a typical object. Because the accretion rate increases with the mass of the accreting object and with the ambient density, which also increases with time as the subgroups condense and merge, the most massive objects tend to undergo a runaway growth in mass.

The most massive object in the simulation of Bonnell and Bate forms in a compact central cluster core which undergoes a runaway increase in density because of dissipative interactions, a result somewhat analogous to the runaway growth of a central singularity in a collapsing isothermal sphere. This most massive object acquires about half of its final mass by accretion and half by mergers, although the importance of mergers may be somewhat exaggerated in this simulation by an artificially large merger cross section. Mergers typically occur in close binary systems when they are perturbed by interaction with another protostar, causing the two protostars in the binary to merge and a new binary to be formed by the capture of the third object. As a result of all of these complex processes, a mass spectrum resembling the Salpeter power law is built up; while there is no simple quantitative explanation for this result, a power law mass spectrum is predicted if the accretion rate increases with a power of the stellar mass. More generally, a power-law mass function might be expected if the accumulation processes that build up the more massive stars are scale-free and involve no new mass scale larger than the Jeans mass. The simulations thus appear to capture, at least qualitatively, many aspects of star formation in clusters, including the formation of the most massive stars at the centre, the accompanying formation of many binary and multiple systems, and the emergence of a realistic stellar mass spectrum. The simulations also demonstrate clearly that these processes are all strongly coupled and that they involve complex dynamics and interactions between protostars and residual gas. The formation of the most massive stars can perhaps be regarded as the culmination of all of the processes.

Formation of the most Massive Objects

A question of long-standing interest is whether there is an upper limit on stellar masses. Although it has been suggested that feedback effects might set an upper limit of the order of 60 $M_\odot$ on the mass that an accreting protostar can attain, these effects depend on the detailed dynamics of the star formation process, and if a protostar grows by accreting extremely dense gas or by mergers, there is no clear upper limit to the mass that it can attain. It is also not clear from observations that there is any well-defined upper limit on stellar masses, since clusters with larger masses tend to contain more massive stars, and stars with masses up to at least 150 $M_\odot$ are found in some young clusters with masses of several times 10^4 $M_\odot$. The mass of the most massive star in a star-forming cloud increases systematically with the mass of the cloud, and on the basis of more limited data, a similar correlation appears to hold between with the mass of the most massive star in a cluster and the mass of the cluster, this relation being approximately $M_{star} \sim 1.2\, M_{cluster}^{0.45}$. Although the data on which this relation is based are sparse, the relation does not clearly terminate at any maximum mass, and this suggests that clusters with even larger masses might form even more massive stars. In this case, it might be that some globular clusters with masses well above 105 $M_\odot$ once contained even more massive stars that have long since disappeared.

If stars with masses significantly above 150 $M_\odot$ are sometimes formed in massive clusters, this could have some important consequences. Metal-poor stars with masses between 140 and 260 $M_\odot$ are predicted to explode completely as very energetic supernovae at the end of their lifetimes, while stars more massive than 260 $M_\odot$ are predicted to collapse completely to black holes. Thus, if stars with masses above 260 $M_\odot$ were ever to form at the centres of very massive clusters, the end result could be the formation of a black hole of similar mass at the cluster centre. In sufficiently dense young clusters, the runaway merging of already formed massive stars might also result in the formation of very massive stars and eventually black holes at the cluster centre. There has been much interest in possible evidence for the existence of massive black holes at the centres of some massive globular clusters, although the evidence is so far not conclusive. In addition, some ultraluminous X-ray sources that are possibly associated with luminous young clusters in other galaxies have been interpreted as 'intermediate-mass black holes' with masses of hundreds to thousands of solar masses. If massive black holes can indeed be formed at the centres of very massive clusters, and if such massive clusters tend to be concentrated near the centres of galaxies, this could be one path towards building up supermassive black holes in galactic nuclei. All of these possibilities remain very speculative, however, and none has yet been supported by additional evidence, so it remains unclear whether exceptionally massive stars and black holes can indeed form by mechanisms like those discussed here. Nevertheless, the importance of understanding the formation of the most massive stars is clear, and an improved understanding of the processes involved may eventually help in understanding the processes involved in the formation and growth of massive black holes.

Stellar Luminosities

The total energy emitted from the surface of a star per unit time (which is the total power of the star) is referred to as its luminosity L. The luminosity of the Sun, for example, is 3.9×10^{26} J/s (or watts). The amount of energy from the Sun that reaches a particular planet depends on the distance of the planet from the Sun, since solar energy is emitted in all directions and spreads throughout space. By the time the solar radiation reaches a distance d from the Sun, it has been spread over an area of $4\pi d^2$, as illustrated:

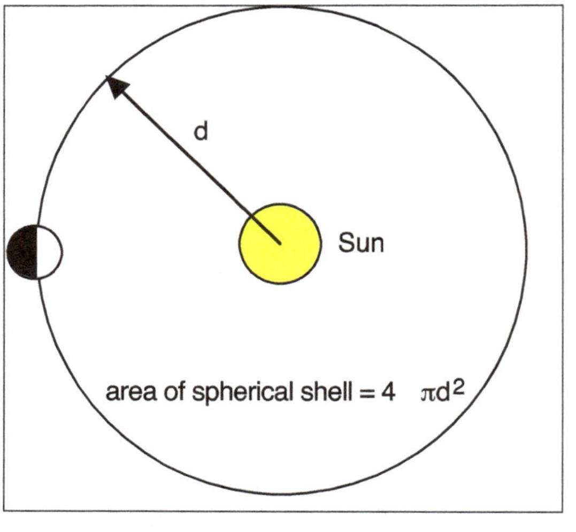

The amount of energy per unit time crossing an element of area facing the Sun, but a distance d away from it, is referred to as the flux F (may also be called the apparent brightness in astronomy). In terms of the luminosity, the flux is given by:

$$F = L / 4\pi d^2$$

and has units of energy per unit area per unit time. Further, there is nothing special about the Sun in this equation, it applies to all stars.

Example: The solar luminosity is 3.9×10^{26} J/s, and the corresponding energy flux from the Sun as seen on the Earth (a distance of 1.5×10^{11} m away) is 1380 watts/m^2.

If one can determine the luminosity of a star without knowing d, then a measurement of the flux F on Earth can be inverted to find d. That is:

- Extract L from some observable characteristic of the star.
- Measure F on Earth.
- Use $F = L / 4\pi d^2$ to solve for d.

The problem with this approach is that dust and gas between Earth and the star in question tend to reduce F and give a calculated distance that is longer than the true distance. One may be able to correct for part of this using the spectrum of light from the star, which will be depleted in the blue because of scattering.

Methods for finding L generally involve selecting a particular observable (e.g. surface temperature) of nearby stars whose distance is known from parallax, then measuring the same observable on a distant star. For example:

- L is related to surface temperature for many stars: Light emitted from a hot object has a distribution of wavelengths that is specific to the object's temperature. For example, most stars have surface temperatures in the 2,500 to 30,000 K range, with the Sun having a surface temperature of 5,800 K. The luminosity of young stars is observed to increase steadily with their temperature, so a measurement of temperature (from wavelength) provides a measure of L. Knowing T by itself doesn't tell us L unless the radius if known.
- L is related to the pulsation period of Cepheid variables: Originally discovered by amateur astronomer John Goodricke in 1794, Cepheids are stars whose luminosity oscillates with periods of roughly 1 to 50 days. With large luminosities in the 102 to 104 times that of the Sun, they can be seen at distances of millions of parsecs, well beyond our galaxy. Astronomer Henrietta Leavitt established that Cepheid luminosity is a unique function of the oscillation period, through her studies of nearby Cepheids.

Apparent Magnitude m

Efforts to quantify the brightness of stars date back to the Greeks. Hipparchus assigned an apparent magnitude m to a selection of visible stars, starting with m = 1 for the brightest stars visible in his part of the world. The apparent magnitude increases as the brightness decreases, with the faintest stars having the largest values of m.

Brightness, according to our eyes or a photometer, is determined by the flux F on Earth, not the luminosity L at the source. Because the response of the human eye is more logarithmic than linear,

$$[magnitude] \propto log[flux]$$

In the modern definition of m, a difference of 5 magnitudes is assigned to a factor of 100 in flux. Thus, if we compare two stars 1 and 2,

$$\frac{F_2}{F_1} = 100^{(m_1 - m_2)/5} = 10^{(m_1 - m_2)/2.5}$$

Taking $\log_{10}$ of both sides,

$$m_1 - m_2 = 2.5\log_{10}(F_2 / F_1)$$

or

$$m_1 - m_2 = 2.5\log_{10}(F_1 / F_2).$$

Aside: the minus sign confirms that m increases as F decreases.

The reference point of m is fixed by convention; the brightest star is Sirius with a negative apparent magnitude of m = -1.43. Many familiar stars have m in the +3 to +8 range. As seen from Earth, the Sun is much brighter than any other star and has m_{sun} = -26.81.

Absolute Magnitude M

Other brightness scales are more useful in certain circumstances. One approach is to define an "absolute magnitude" M that eliminates the distance dependence of the apparent magnitude:

M = m if the observer were 10 pc from the star

M is a better reflection of the luminosity.

Imagine two observers measuring a distant star:

- From Earth: m and F.
- From 10 pc: M and F_{10}.

Thus,

$$100^{(m-M)/5} = \frac{F_{10}}{F} = \left(\frac{d}{10\,\text{pc}}\right)^2$$

where the latter equality follows from $F = L / 4\pi d^2$. Taking logarithms again,

$$2(m-M)/5 = 2\log_{10}(d/10pc)$$

or

$$m-M = 5\log_{10}(d/10pc).$$

Stellar Radii

Although the structure of galaxies can be seen with a telescope, stars appear to be point-like. What techniques are available to us for determining their sizes? Here, we concentrate on two approaches - the Hanbury-Brown Twiss effect, and stellar luminosities.

Hanbury-Brown/Twiss

Photons are bosons, which mean that two of them can occupy the same state. This is in contrast to fermions, no two of which can share the same quantum numbers. In a thermal distribution of particles, we will show later that compared to a distribution of classical particles (i.e., ignoring the effects of spin), at low relative momenta:

- Bosons pairs have a higher probability of occurrence.
- Fermion pairs have a lower probability of occurrence.

Consider, then, pairs of photons emitted from a star and observed some time later on Earth. Defining as 1 the number of photon pairs in a random distribution having a given relative momentum Δp, we expect

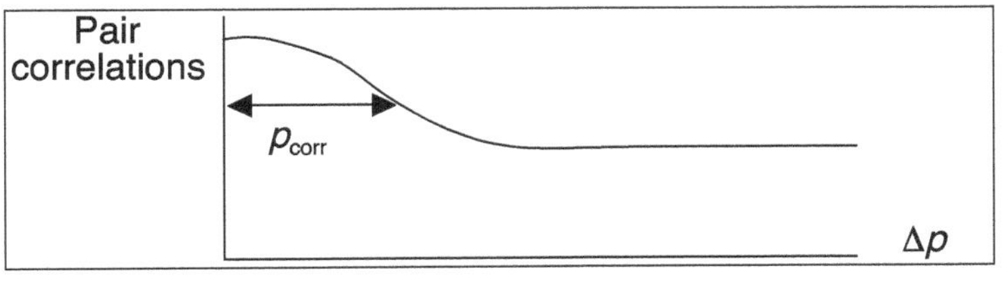

Crudely speaking, the region of momentum p_{corr} over which the pairs are enhanced is related to the size of the star by,

$$R \sim h / p_{corr}.$$

Only the brightest stars can be analyzed with this technique. To date, more than 30 stars in the southern hemisphere have been analyzed, using an interferometer in Australia.

Luminosity and Stephan-Boltzmann

This technique is a little less direct, and its implementation requires several pieces of data. The strategy is to measure the star's surface temperature T, luminosity L and then use the Stephan-Boltzmann equation for the flux $F_{surf} = \sigma T^4$ to find the star's radius via $L = 4\pi R^2 F_{surf}$.

That is,

$$R = \left[\frac{L}{4\pi\sigma T^4} \right]^{1/2}$$

where σ = Stefan-Boltzmann constant = 5.67 x 10^{-8} watts / m^2K^4.

Example:

Sirius is the brightest star, and is actually a binary. Located just 2.6 pc from Earth, its distance is accurately known from parallax. Thus, measuring the flux of its light at the Earth allows the determination of its luminosity from $4\pi d^2 F_{at\ Earth}$. The surface temperature can be obtained from the spectrum of light from each member of the binary pair. The resulting values are:

$$\begin{array}{llll} \textit{Sirius A} & L = 23.5 \text{ solar luminosity} & T = 9910\ K \\ \textit{Sirius B} & L = 0.03 \text{ solar luminosity} & T = 27000\ K \end{array}$$

where the solar luminosity is 3.9 x 10^{26} watts. Let's find the radii of these two stars:

Sirius A:

$$R = \left[\frac{23.5 \bullet 3.9 \times 10^{26}}{4\pi \bullet 5.67 \times 10^{-8} \left(9910\right)^4} \right]^{1/2} = 1.15 \times 10^9\ m$$

Sirius B:

$$R = \left[\frac{23.5 \bullet 3.9 \times 10^{26}}{4\pi \bullet 5.67 \times 10^{-8} \left(27000\right)^4} \right]^{1/2} = 1.15 \times 10^6\ m$$

Compared to the radius of the Sun at 6.960 x 10^8 m, Sirius A is twice as large. Sirius B is a hundred times smaller!

In fact, Sirius B is slightly smaller than the Earth, whose radius is 6400 km. Initially resolved as a star in 1862, the temperature of Sirius B was determined in 1915, demonstrating that it was completely different from conventional stars – the first example of a white dwarf.

Chemical Composition of Stars

Our galaxy, the Milky Way, is home to over 400 billion stars of varying brightness. The majority of these stars are described as being main sequence, which means their cores are fusing hydrogen to create helium. The Sun is a main sequence star and its chemical composition mainly consists of hydrogen and helium with trace amounts of other elements.

Hydrogen

Hydrogen is the most abundant element in the universe and makes up three-quarters of all matter. Stars form when huge amounts of gas and dust collapse under their own gravitational force. The majority of this gas is hydrogen which is the basic fuel that stars use to create energy. During hydrogen fusion, protons (nuclear subatomic particles) are combined in order to create helium. Other by-products are also created in this reaction such as electrons, positrons (antielectron), gamma rays and neutrinos. Neutrinos are ghost like particles that do not interact strongly with matter so these usually escape from the Sun. The collision of the remaining particles with surrounding atoms leads to the heating of the Sun.

Helium

Helium is the second most abundant element in the universe and is a major component of main sequence stars such as the Sun. Helium accumulates in the core of stars as a result of hydrogen nuclear fusion. Helium accounts of approximately 27 percent of the Sun's mass.

Carbon

When hydrogen levels within a star's core depletes, the standard fusion reaction can no longer take place. This leads to a decrease in the amount of energy radiating outwards and the stellar core collapses increasing the temperature and pressure. When the temperature reaches 200 million Kelvin, helium fusion becomes possible. Three helium nuclei fuse to create a single carbon atom.

Oxygen and other Trace Elements

Fusion of four helium nuclei can be used to create oxygen atoms. This happens in stars that have used up their supply of hydrogen within the core. Further fusion processes can create heavier elements such as silicon, magnesium and sodium. However, the abundance of these elements in most stars is very low and accounts for less than 1 percent of the mass. Fusion within stars can only account for the creation of elements up to the mass of iron. Beyond this, the fusion process uses energy rather than creates it. The remaining heavy elements beyond iron are thought to be forged in the collapse of heavy stars -- a process known as supernova.

Stellar Opacity

The study of stellar opacity has long been regarded by astrophysics as a rather unattractive pursuit from a number of different standpoints.

The mean absorption coefficient, κ, is not a constant; it is dependent on frequency, and is therefore frequently written as κ_ν. Inside a star, several different sources of opacity are important, and each has its own frequency dependence. To compute the appropriate "average" opacity, $\bar{k}$ recall that Fick's law states that the flux at any point is related to the energy density U by,

$$F_\nu = -D\nabla U = -\frac{1}{3}\bar{v}l\nabla n = -\frac{1}{3}\frac{c}{kvp}\nabla U$$

and that, when integrated over all energies,

$$F = -\frac{4acT^3}{3\bar{k}p}\frac{dT}{dr}$$

But how is $\bar{k}$ related to κ_ν? To answer this question, we start by explicitly writing in the frequency dependence of U, i.e.,

$$U_\nu = -\frac{4\pi}{c}B_\nu(T) = \frac{4\pi}{c}\frac{2h\nu^3}{c^2}\left\{\frac{1}{e^{h\nu/kT}-1}\right\}$$

and compute the monochromatic flux distribution,

$$F_\nu = -D\nabla U_\nu = \frac{c}{3p}\frac{1}{k_\nu}\nabla\left\{\frac{4\pi}{c}B_\nu(T)\right\} = -\frac{4\pi}{3p}\frac{1}{k_\nu}\frac{dB_\nu}{dT}\frac{dT}{dr}$$

The total flux, integrated over all frequencies is then,

$$F = -\frac{4\pi}{3p}\frac{dT}{dr}\int_0^\infty \frac{1}{k_\nu}\frac{dB_\nu}{dT}d\nu$$

Comparing equation $F = -\frac{4\pi}{3p}\frac{dT}{dr}\int_0^\infty \frac{1}{k_\nu}\frac{dB_\nu}{dT}d\nu$ and $F = -\frac{4acT^3}{3\bar{k}p}\frac{dT}{dr}$, we see that,

$$\frac{1}{\bar{k}} = -\frac{\pi}{acT^3}\int_0^\infty \frac{1}{k_\nu}\frac{dB_\nu}{dT}d\nu$$

or, since

$$\int_0^\infty \frac{dB_\nu}{dT}d\nu = \frac{d}{dT}\int_0^\infty B_\nu d\nu = \frac{dB}{dT} = \frac{ac}{\pi}T^3$$

$$\frac{1}{\bar{k}} = \frac{\int_0^\infty \frac{1}{k_\nu}\frac{dB_\nu}{dT}d\nu}{\int_0^\infty \frac{dB_\nu}{dT}d\nu}$$

The average opacity that you should use is therefore that which is harmonically weighted over the temperature derivative of the Planck curve. This is called the Rosseland mean opacity. Note that the function weights the high frequencies more than the low frequencies. To see this, we can take the temperature derivative of the Planck function,

$$\frac{dB}{dT} = \frac{2h^2\nu^4}{c^2kT^2}\frac{e^{h\nu/kT}}{\left\{e^{h\nu\ kT}-1\right\}^2} = \left(\frac{2k^3T^2}{c^2h^2}\right)\left(\frac{x^4e^x}{\left\{e^x-1\right\}^2}\right)$$

where x = hν/kT, and find its peak by taking the derivative. Then,

$$\left(\frac{2k^3T^2}{c^2h^2}\right)\left\{\frac{x^4e^x+4x^3e^x}{\left(e^x-1\right)^2}-\frac{2x^4e^{2x}}{\left(e^x-1\right)^3}\right\}=0$$

$$x^3e^x\left(e^x-1\right)^{-3}\left(xe^x-x+4e^x-4-2xe^x\right)=0$$

$$x^3e^x\left(e^x-1\right)^{-3}\left(4e^x-4-x-xe^x\right)=0$$

Mathematically, this works out to x ≈ 3.83. In other words, the high energy photons count much more than low energy photons.

Although in real models, the Rosseland mean opacity is computed numerically, it is instructive to evaluate the form of $\bar{k}$ when $\kappa_\nu \propto \nu^{-n}$. From the definition of the Rosseland mean,

$$\frac{1}{k} \alpha \frac{\frac{2h^2}{c^2kT^2}\int_0^\infty \nu^{n+4}\frac{e^{h\nu/kT}}{\left(e^{h\nu/kT}-1\right)^2}d\nu}{\frac{2h^2}{c^2kT^2}\int_0^\infty \nu^{4}\frac{e^{h\nu/kT}}{\left(e^{h\nu/kT}-1\right)^2}d\nu}$$

which, in terms of x is,

$$\frac{1}{k} \alpha \frac{\int_0^\infty \left(\frac{kT}{h}\right)^{n+4} x^{n+4}\frac{e^x}{\left(e^x-1\right)^2}\left(\frac{kT}{h}\right)dx}{\int_0^\infty \left(\frac{kT}{h}\right)^{4} x^{4}\frac{e^x}{\left(e^x-1\right)^2}\left(\frac{kT}{h}\right)dx}$$

Or

$$\frac{1}{k} \alpha \left(\frac{kT}{h}\right)^n \frac{\int_0^\infty x^{n+4}\frac{e^x}{\left(e^x-1\right)^2}dx}{\int_0^\infty x^{4}\frac{e^x}{\left(e^x-1\right)^2}dx} \alpha T^n$$

Thus, if the opacity law is $\kappa \propto \nu^{-n}$, the temperature dependence of the Rosseland mean opacity will be $\bar{k} \propto T^{-n}$. This follows since the mean opacity must drop as the number of high energy photons increases.

Stellar Temperatures

A Star's Surface Temperature can be Determined from its Spectrum

The Sun produces a spectrum which is a close approximation to a blackbody spectrum; so do other stars. The temperature of a blackbody is given by a relatively simple formula:

$$T = 0.0029/\lambda_{max},$$

where T = temperature of the blackbody (measured in degrees Kelvin)

and λ_{max} = wavelength of maximum emission (measured in meters).

Using the above formula, we can compute the temperature of a star's photosphere from the wavelength at which it emits the maximum amount of light.

The temperature of a star's photosphere can also be deduced from its color. Cool stars (such as Betelgeuse, which has a surface temperature of T = 3500 Kelvin) emit more red and orange light than blue and violet light. Thus, cool stars are red. Hot stars (such as Rigel, which has a surface temperature of T = 15,000 Kelvin) emit more blue and violet light than red and orange light. Thus, hot stars are blue.

One way of classifying stars is by their temperature; stellar temperatures run from about 2500 Kelvin to about 50,000 Kelvin. Another way of classifying stars is by using the notorious OBAFGKM spectral classes. They are 'notorious' because the ordering of the spectral classes, from "O" to "B" to "A" to "F" to "G" to "K" to "M", doesn't follow in logical alphabetical order. Initially, in the 19th century, the spectral classes represented a purely empirical method of sorting stellar spectra, based on the strength of their hydrogen absorption lines.

- "A" stars had the strongest absorption lines.
- "B" stars had the next strongest, and so on through the alphabet.

In the twentieth century, when the physics of absorption lines was better understood, it was realized that spectral class represents a particular temperature range. The "O" stars are the hottest, and the "M" stars are the coolest.

- O= 40,000 Kelvin

- B = 20,000 Kelvin
- A = 9000 Kelvin
- F = 7000 Kelvin
- G = 5500 Kelvin
- K = 4500 Kelvin
- M = 3000 Kelvin.

The OBAFGKM spectral sequence has recently been extended to include class L (objects with temperature around 2000 Kelvin) and class T (with temperature less than 1300 Kelvin). Objects of spectral type L and T are not (technically speaking) stars at all, since they are not hot enough for fusion to occur in their cores. They are called "brown dwarfs" rather than "stars".

The Hertzsprung-Russell Diagram Sorts Stars by Luminosity and Temperature

Temperature doesn't tell the whole story. For instance, Betelgeuse and Proxima Centauri have the same surface temperature (they are both M stars). However, Betelgeuse is 300 million times more luminous.

Another way of classifying stars is by luminosity.

- Most luminous stars: L = 1,000,000 L_{sun}
- Dimmest stars: L less than L_{sun}/10,000

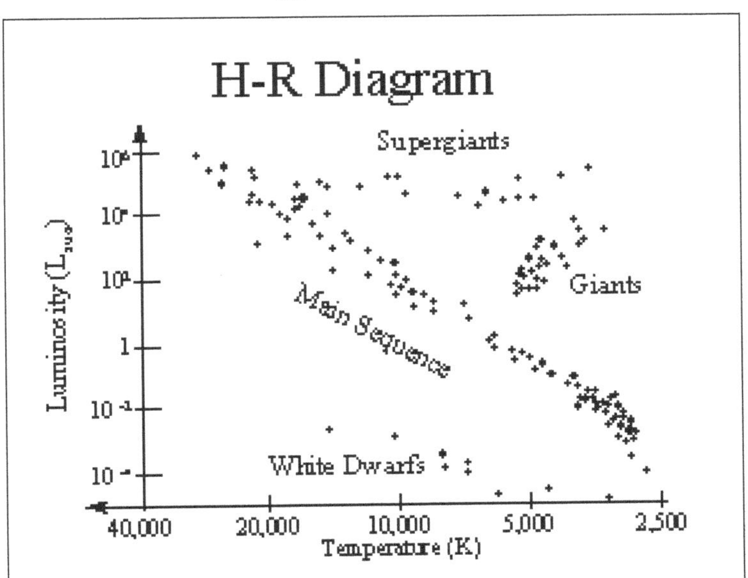

Just prior to WWI, Ejnar Hertzsprung and Henry Norris Russell separately had the same idea - classify the stars according to both their temperature and their luminosity.

Make a plot - temperature on the horizontal axis (hot stars to the left; cool stars to the right) and luminosity on the vertical axis (dim stars at the bottom; bright stars at the top). This plot is called the Hertzsprung-Russell diagram (or H-R diagram, for short) after its inventors. An example of a Hertzsprung-Russell diagram is given below.

H-R diagrams are very useful tools, so we'll be referring to them throughout the entire first half of the course.

Most Stars are on the Main Sequence of the Hertzsprung-Russell Diagram

Stars are not randomly distributed across the H-R diagram.

About 90% of all stars are on a narrow diagonal band running from the upper left corner of the H-R diagram (hot, luminous stars) to the lower right corner (cool, dim stars). This diagonal band is called the main sequence.

The Sun is on the main sequence.

- All main sequence stars hotter than the Sun are more luminous than the Sun and larger than the Sun (Example: Sirius). The hottest main sequence stars are "O" stars with T = 40,000 K, L = 300,000 L_{sun}.
- All main sequence stars cooler than the Sun are less luminous than the Sun and smaller than the Sun (Example: Proxima Centauri). The coolest main sequence stars are "M" stars with T = 2000 K, L = 0.00005 L_{sun}.

The above statements are true only for main sequence stars. The remaining 10% of stars don't fall on the main sequence.

Giants are more luminous than a main sequence star of the same temperature. Giants tend to be relatively cool (T < 6000 Kelvin) but luminous (L = 100 to 1000 L_{sun}). Giants are rare, but easy to detect, because of their high luminosity.

Supergiants are even more luminous than giants. Supergiants can have any temperature, but they are always very luminous, with L = 100,000 to 1,000,000 L_{sun}. Supergiants are very rare, but are very easy to detect, because of their very high luminosity.

White Dwarfs are less luminous than a main sequence star of the same temperature. They are called white dwarfs because they are fairly hot; white-hot, in fact, with temperatures of T > 5000 Kelvin. They are low in luminosity, with L = 0.0001 to 0.01 L_{sun}. White dwarfs are fairly common, but they are difficult to detect, because of their low luminosity.

Stellar Magnetic Fields

Stellar magnetic fields are an array of forces that can be observed surrounding and at the surfaces of stars like the sun. They are similar in nature to the effect of the well–known dipolar magnets found in science laboratories, classrooms, and toys, but far more powerful and infinitely more complex. They are an important part of the physical makeup of stars because they affect their interiors, atmospheres, and immediate environments. Observations of the sun show that it has a dynamic overall magnetic field and also smaller, but often much stronger pockets of magnetism associated with sunspots. The influence of these more localized magnetic fields can sometimes be quite dramatic when they are involved in the creation, shaping, and size of solar prominences, flares, and some features in the solar atmosphere (the corona). The large–scale magnetic field of the sun helps determine processes by which chemical elements are transported within and around the sun, and even the spin, or rotation, of the stellar surface. The study of the sun's magnetic fields, particularly their large–and small–scale structures, helps in understanding their origins. It is assumed by astronomers that when magnetic fields around stars other than the sun are studied in detail they will show similar features and dynamics. Knowledge gained about the magnetic fields of stars can lead to an understanding of their potential impact on long–term stellar evolution.

Exactly how stellar magnetic fields work is somewhat of a mystery. The most widely accepted explanation for them is called the dynamo model. The dynamo principle is used in generators on the Earth, but may be thought of as the reverse of what is happening in a star. In a simple emergency generator, a gas engine spins a magnet within a coil of wire. The interaction of the moving magnetic field within the coil generates electricity in the wire, which is then sent out to a connector that provides electrical power to devices outside the generator. This is how hydroelectric power is generated at dams like the famous Hoover Dam in the Black Canyon of the Colorado River (between Arizona and Nevada). However, instead of a gas engine, water under high pressure provides the motion required to make the generator work.

In a stellar dynamo, rather than electricity being generated because of a moving magnetic field, a magnetic field seems to be generated by two major motions within the star. The first motion is the movement of the gases in the convection zone, which makes up the upper layer of the star. In this region, material at and just beneath the surface moves up as heat is transferred outward from lower layers to the surface by a process in which hot gas rises just as hot air does on the Earth. Once some of the heat of the gas is released at the surface of the Sun, that gas drops down again as it is replaced by hotter gases from below.

The second motion is caused by the simple fact that the sun is made of gas. Because of this, it does not rotate at the same speed everywhere as would a solid object like a planet. This is called differential rotation and it causes the material at the equator to move faster than material at the poles. While scientists have not worked out all the details, it appears that these two effects together create the basic stellar magnetic field of the sun and other stars. However, to be able to create a full picture, it would be necessary to describe accurately all the physical processes operating on the surface of and in the interior of every area of the sun including small– and large–scale turbulence. In addition, the overall magnetic field and sunspot fields themselves effect the movements of the convection zone, creating a situation far more complex than the highly unpredictable weather patterns of the Earth. A deeper understanding of the causes of stellar magnetic fields will require observations of many more stars and a more complete understanding of processes within them.

On the sun, more localized magnetic fields can be found and are made visually obvious by the appearance of sunspots, which were first recorded by ancient Chinese astronomers. They can be so large that they can indeed be observed, with proper filtering, with the naked eye. In the 1600s, Italian astronomer and physicist Galileo Galilei and his contemporaries rediscovered sunspots shortly after the start of telescopic astronomy (astronomy that was supported by the use of telescopes). Sunspots are regions on the solar surface that appear dark because they are cooler than the surrounding surface area (photosphere) by about 2,200 °F (1,200 °C). This means they are still at a temperature of about 7,600 °F (4,200 °C). Even though they look dark in photographs of the sun, they are still very bright. If a piece of sunspot could be brought to the Earth, it would be extremely hot and blinding to look at just as any other piece of the sun. Sunspots develop and persist for periods ranging from hours to months, and are carried around the surface of the Sun by its rotation. Sunspots usually appear in pairs or groups and consist of a dark central region called the umbra and a slightly lighter surrounding region called the penumbra. The rotation period of the sun was first measured by tracking sunspots as they appeared to move around the sun. Galileo used this method to deduce that the sun had a rotational period of about one month. However, because the sun is not a solid body, it does not have one simple rotational period. Modern measurements indicate that the rotation period of the sun is about 25 days near its equator, 28 days at 40° latitude, and 36 days near the poles. The rotation direction is the same as the motion of the planets in their orbits around the sun.

The magnetic causes of sunspots were not known until the early years of the twentieth century when American solar astronomer George Ellery Hale mapped the solar magnetic field through its effect—called the Zeeman effect—on the detailed shape and polarization of spectral lines. Spectra show the chemical makeup of stars and are a major source of information for astronomers. They are created by spreading the light of a star into its component parts in the same way a prism creates a rainbow of colors from a light source. Chemical reactions in the star create lines of different intensities at predictable places along the spectrum allowing scientists to determine the makeup of the star. Since the chemical reactions would produce a spectral line in a given way in the absence of a magnetic field, scientists can see the effects of fields by comparison to the known spectrum of the reaction. The Zeeman effect is a change in the spectral lines caused by the sun's magnetic field. The sun's magnetic field has been mapped on a regular basis ever since Hale first did it, and it is now known that the 11–year sunspot cycle is just a part of an overall 22–year magnetic cycle. The shape of the sun's magnetic field changes throughout the 11–year cycle, when during this period it reverses its magnetic polarity and begins the whole process over again. In addition to the differential rotation helping to cause the magnetic field of the Sun, it also stretches the north–south magnetic field lines until they run east–west during the first 11 years of the magnetic cycle. Rotating convection then somehow regenerates the north–south field, but with a reversed polarity, causing the process to start again for another 11 years. During these half–cycles, the number and intensity of sunspots increases and decreases with the changes in the overall magnetic field.

Using special high–resolution spectropolarimeters combined with other techniques, the magnetic fields of stars beyond the sun can be detected through the effect they have on the Zeeman signatures found in the shape and polarization state of spectral lines of those stars. Zeeman–Doppler imaging (ZDI) works best for moderate to ultra–fast rotating stars, for which the polarization of individual magnetic regions match the different speeds at which the surface of the star rotates. This method was used to detect the magnetic fields in cool stars other than the sun, showing that the same type of phenomena occur on other stars. Using Zeeman–Doppler imaging, astronomers have managed to detect and map the surface magnetic field of a few extremely active stars of about one solar mass (with ages ranging from a few million to more than ten billion years—twice the sun's age). Some major differences were found between the alignment of the magnetic field lines of these stars and those of the sun, adding to the mystery of understanding stellar magnetic fields. The conclusion of astronomers studying these stars results is that the entire convection zone of these active stars is involved in forming the magnetic field rather than just the upper layers as appears to be the case with the sun.

These methods allow monitoring of the long–term evolution of the magnetic field shape and strength of other stars. Using them, astronomers hope to be able to detect the polarity switch of the large–scale field and observe a stellar analog of the solar magnetic cycle. If a change in magnetic field polarity is observed, it may indicate the approach of a polarity switch in the magnetic field of the star. Such observations would show that stellar magnetic fields are indeed very similar to those of the sun.

Stellar Nucleosynthesis

Stellar nucleosynthesis is the process by which elements are created within stars by combining the protons and neutrons together from the nuclei of lighter elements. All of the atoms in the universe began as hydrogen. Fusion inside stars transforms hydrogen into helium, heat, and radiation. Heavier elements are created in different types of stars as they die or explode.

The idea that stars fuse together the atoms of light elements was first proposed in the 1920s, by Einstein's strong supporter Arthur Eddington. However, the real credit for developing it into a coherent theory is given to Fred Hoyle's work in the aftermath of World War II. Hoyle's theory contained some significant differences from the current theory, most notably that he did not believe in the big bang theory but instead that hydrogen was continually being created within our universe. (This alternative theory was called a steady state theory and fell out of favor when the cosmic microwave background radiation was detected).

The simplest type of atom in the universe is a hydrogen atom, which contains a single proton in the nucleus (possibly with some neutrons hanging out, as well) with electrons circling that nucleus. These protons are now believed to have formed when the incredibly high energy quark-gluon plasma of the very early universe lost enough energy that quarks began bonding together to form protons (and other hadrons, like neutrons). Hydrogen formed pretty much instantly and even helium (with nuclei containing 2 protons) formed in relatively short order (part of a process referred to as Big Bang nucleosynthesis).

As this hydrogen and helium began to form in the early universe, there were some areas where it was denser than in others. Gravity took over and eventually these atoms were pulled together into massive clouds gas in the vastness of space. Once these clouds became large enough, they were drawn together by gravity with enough force to actually cause the atomic nuclei to fuse, in a process called nuclear fusion. The result of this fusion process is that the two one-proton atoms have now formed a single two-proton atom. In other words, two hydrogen atoms have begun one single helium atom. The energy released during this process is what causes the sun (or any other star, for that matter) to burn.

It takes nearly 10 million years to burn through the hydrogen and then things heat up and the helium begins fusing. Stellar nucleosynthesis continues to create heavier and heavier elements until you end up with iron.

Creating the Heavier Elements

The burning of helium to produce heavier elements then continues for about 1 million years. Largely, it is fused into carbon via the triple-alpha process in which three helium-4 nuclei (alpha particles) are transformed. The alpha process then combines helium with carbon to produce heavier elements, but only those with an even number of protons. The combinations go in this order:

- Carbon plus helium produces oxygen.
- Oxygen plus helium produces neon.
- Neon plus helium produces magnesium.
- Magnesium plus helium produces silicon.
- Silicon plus helium produces sulfur.

- Sulfur plus helium produces argon.
- Argon plus helium produces calcium.
- Calcium plus helium produces titanium.
- Titanium plus helium produces chromium.
- Chromium plus helium produces iron.

Other fusion pathways create the elements with odd numbers of protons. Iron has such a tightly bound nucleus that there isn't further fusion once that point is reached. Without the heat of fusion, the star collapses and explodes in a shockwave.

Physicist Lawrence Krauss notes that it takes 100,000 years for the carbon to burn into oxygen, 10,000 years for the oxygen to burn into silicon, and one day for the silicon to burn into iron and herald the collapse of the star.

Nova Spectra

Novae are interacting binary systems with a white dwarf primary and a main sequence K-M type secondary. The white dwarf is generally accepted to be a carbon-oxygen (CO) or an oxygen-neon (ONe) white dwarf, and the main sequence secondary is filling its Roche-lobe. The white dwarf accretes hydrogen rich matter from the secondary as a result of mass transfer via Roche-lobe overflow. The accreted matter forms and accretion disc around the white dwarf. As more matter is constantly accreted, the semi-degenerate nature of the white dwarf surface causes a buildup of pressure and temperature at the base of the accretion disc. At a critical temperature and pressure, hydrogen burning sets in, which soon builds up to a thermonuclear runaway (TNR) reaction that releases a huge amount of energy. The energy released is imparted to the accretion disc, causing the disc to expand and be ejected from the system. A nova explosion thus occurs, accompanied by the release of energy in the range $10^{38} - 10^{40}$ ergs, and ejection of matter with velocities > 500 km s−1. Matter is ejected either in the form of an optically thick wind, as discrete shells, or as a combination of both.

Nova explosions are observed as a sudden brightening of the star by several magnitudes, followed by a decline. A nova outburst leads to the ejection of the accreted matter alone, without disrupting the binary system, leading to a resumption of the accretion process. Repeated outbursts are hence possible in these binary systems. Those systems that have had only one recorded nova outburst are termed as Classical Novae, while the systems that have recorded more than one nova outburst with a recurrence period ranging from a few years to a few decades are termed as Recurrent Novae.

Novae are classified into different speed classes depending on the time taken to decline by two magnitudes from outburst maximum (t_2), or by three magnitudes from outburst maximum (t_3). The various speed classes are very fast (t_2 < 10 days); fast (t_2 = 11 – 25 days); moderately fast (t_2 = 26 – 80 days); slow ($_{t2}$ = 81 – 150 days) and very slow (t_2 = 150 – 250 days).

The Outburst Spectrum

The spectrum of a nova system, following its outburst, shows clear signatures of an expanding material, as well as a TNR reaction and is influenced by the evolution of the photosphere, wind, and surface nuclear reactions, all of which are related. During the initial stages, when the ionization levels are low, the spectrum is dominated by permitted, recombination lines. The ionization levels increase with time as the layers closer to the ionizing source (central WD) are revealed as the ejecta expand. Forbidden and high ionization emission lines are seen at this stage. As the nova approaches its post-outburst quiescence phase, the ionization levels decrease once again.

Evolution of Outburst Spectra

The evolution of the outburst spectrum broadly follows the light curve evolution, as was first described in detail by McLaughlin. The evolution from the premaximum all the way to the late phases are described below:

- Pre-maximum spectrum: The spectrum during the pre-maximum phase is that of an optically thick, cooling ejecta, with blue shifted absorption lines indicating expansion of the material.
- Principal spectrum ($\Delta m \sim 0.6$): The principal spectrum occurs close to visual maximum. At maximum, the spectrum is characterized by strong absorption lines and resembles A-F supergiants with enhanced CNO lines. The aborption lines indicate velocities that are larger than that seen during the pre-maximum phase, and are correlated with the speed class. At, or immediately after maximum, an emission-line component appears in the principal spectrum. The strongest lines are due to H, Ca II, Na II, Fe II, N, He and O.
- Diffuse enhanced spectrum ($\Delta m \sim 1.2$): This is the third absorption system, with broad diffuse absorption lines of species similar to those in the principal system, but with velocities that are twice those of the principal system. The absorptions reach a maximum at $\Delta m \sim 2$, when the lines show P-Cygni profiles with broad emissions of the diffuse enhanced system underlying those of the principal spectrum. In the later phases of this stage, the lines often split into narrow components.
- Orion system ($\Delta m \sim 2.1$): The nova spectrum, which is now a mixture of the principal and diffuse enhanced system, is further complicated by the presence of yet another absorption system, the Orion system. The absorption lines are

diffuse, with velocities that are atleast as much as the diffuse enhanced system. The absorption systems remain diffuse until they disappear at $\Delta m \sim 3.2$, when the nova enters the transition phase. The maximum of this system occurs at $\Delta m \sim 2.7$. The excitation and ionization levels increase with time all through the Orion system, and the emission line component also increases in strength as the absorption component disappears. The lines are generally structured.

- Nebular and post-nova phase ($\Delta m \sim 4.4$): The nebular phase is the final stage of the nova outburst before the nova enters its post-nova quiescence phase. This phase is marked by the presence of high ionization lines such as the auroral, nebular forbidden lines, and high excitation coronal lines. The high excitation lines gradually fade and in the post-nova phase the spectrum is dominated by that of the accretion disc.

Spectral Classification

The spectra of different novae appear quite diverse. However, Williams grouped them based on common characteristics in this seemingly diverse ensemble. The classification of novae is based on their outburst spectra in the optical. All novae are dominated by the strong hydrogen Balmer lines. Hence, classification is best done using the strongest non-Balmer lines present in the optical spectrum during the first few days following an outburst. It is found that all novae spectra either have lines due to Fe II or due to He/N as the strongest non-Balmer lines during the early phases.

Hence, novae can be classified into two broad groups, the "Fe II " and the "He/N" class. However, there are a few novae that fall in between the two classes and are termed "hybrid" novae.

Fe II Class

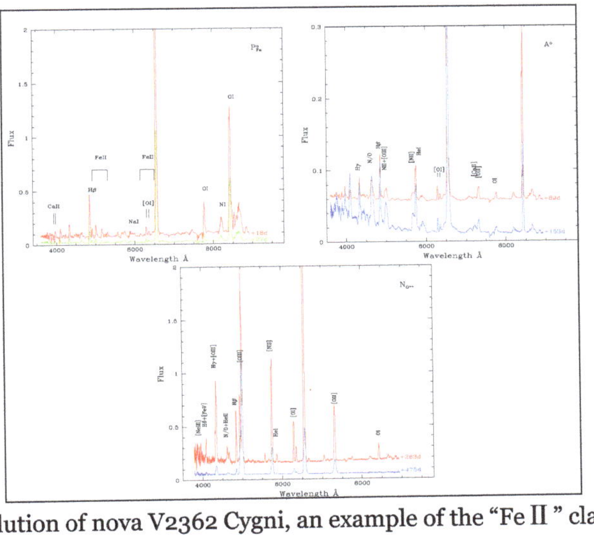

Spectral evolution of nova V2362 Cygni, an example of the "Fe II " class of novae.

The strongest non-Balmer lines during the early phases in this class are due to Fe II. The early phase spectra show pronounced P-Cygni absorption, with velocities < 2500 km s^{-1}. Low ionization transitions are seen during this phase. Novae belonging to the Fe II class are generally the moderately fast to the slow novae, with the spectra evolving over timescales of weeks. The early nebular phase spectrum is dominated by low ionization auroral lines, while some Fe II novae develop strong [Ne II] lines.

The dominant ejection mechanism for this class of novae appears to be wind ejection. Figure shows the spectral evolution of nova V2362 Cygni that belongs to the "Fe II " class.

He/N class

The strongest non-Balmer lines in this class are either the helium (He I/He II) or the nitrogen (N II /N III) lines. The excitation levels in this class of novae are found to be higher than the "Fe II " class, and P-Cygni absorptions are either weak or absent. The lines are generally broad, indicating high expansion velocities, with the velocity being as high as 10,000 km s^{-1} in some cases. The line profiles are boxy and structured indicating shell ejection. The "He/N" novae generally belong to the very fast and fast class of novae, and also show a faster spectral evolution.

The nebular phase spectrum is quite diverse. Some novae develop high excitation coronal lines, while some do not show any forbidden lines during the nebular phase. Yet other novae show strong Ne III forbidden lines. These are also termed as Ne novae and are presumed to have a ONe white dwarf. Figure shows the spectral evolution of the He/N nova V2491 Cygni.

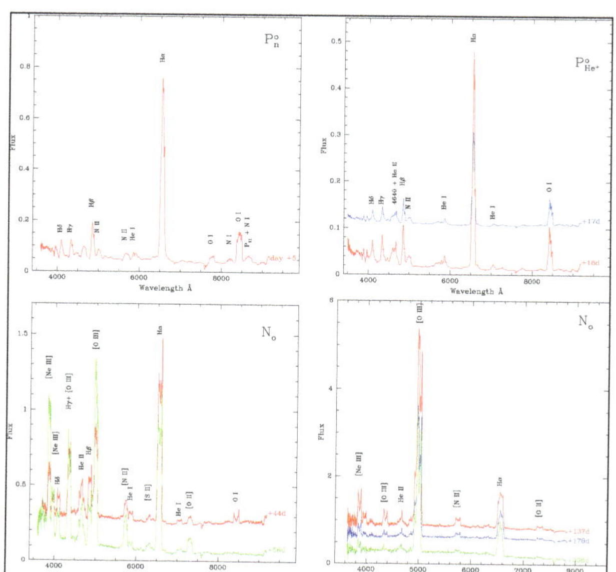

Spectral evolution of nova V2491 Cygni, an example of the "He/N" class of novae. Note the faster evolution compared to the "Fe II " class.

Hybrid Novae

The hybrid novae are those systems that change their spectral class from "Fe II" to "He/N" during the early permitted lines phase, before the forbidden lines appear. Some novae belonging to this class show evidence for simultaneous emission from both components. The line widths indicate higher velocity compared to the "Fe II" class. The line profiles indicate ejection of matter via a weak wind as well as shell ejection. Figure shows an example of a hybrid nova.

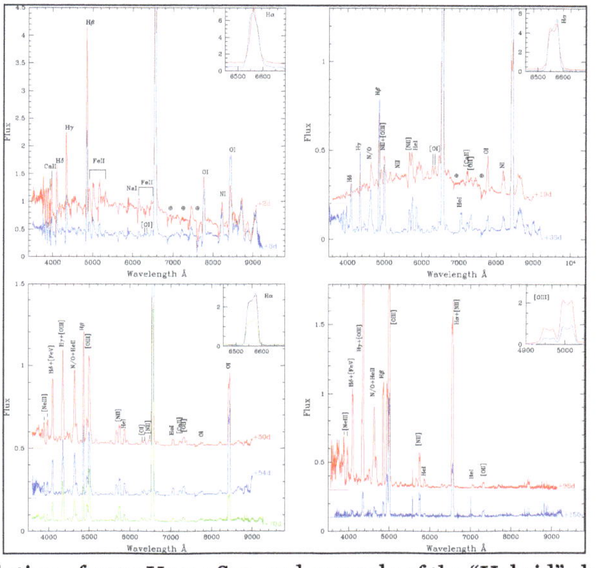

Spectral evolution of nova V5114 Sgr, and example of the "Hybrid" class of novae. Note the change from an "Fe II" class to "He/N" class during the early phases.

The Tololo Classification Scheme

Williams et al. devised a physical basis to describe the outburst spectral evolution, generally referred to as the Tololo Classification Scheme. This scheme defines phases in evolution in terms of the characteristics of the stronger emission lines. These phases are related to the physical conditions in the shell through the ionizing radiation field and mean gas density. Hence, the classification provides an insight to the physical conditions in the nova ejecta as the outburst evolves with time. The main points describing this scheme are:

- The spectrum is classified as belonging to either phase C, P, A, or N.
- Phase C (coronal): A spectrum is considered to be in the coronal phase C if, irrespective of all other line strengths, [Fe X] 6374 Å line is present and stronger than [Fe VII] 6087.
- Phase P (permitted): If the spectrum is not in phase C, then it can be classified as in the permitted phase P if the strongest non-Balmer line is a permitted line.

- The species of this permitted line is mentioned in the subscript. For example, in the case of "Fe II" novae, this phase would be represented by Pfe, while in the case of "He/N" novae this phase would be Phe, or PN.
- Phase A (auroral): If the spectrum is not in phase C, it is considered to be in the auroral line phase whenever any auroral forbidden line has flux higher than the strongest non-Balmer permitted lines, irrespective of the strengths of other nebular lines.
- Phase N (nebular): The spectrum is considered to be in the nebular phase N if it is in neither phase C nor phase A, and the strongest non-Balmer line is a forbidden nebular transition.
- If the OI 8446 Å line is present and prominent, it is represented an 'o' in the superscript, irrespective of the phase.

Supernovae

Supernova, plural supernovae or supernovas, is any of a class of violently exploding stars whose luminosity after eruption suddenly increases many millions of times its normal level.

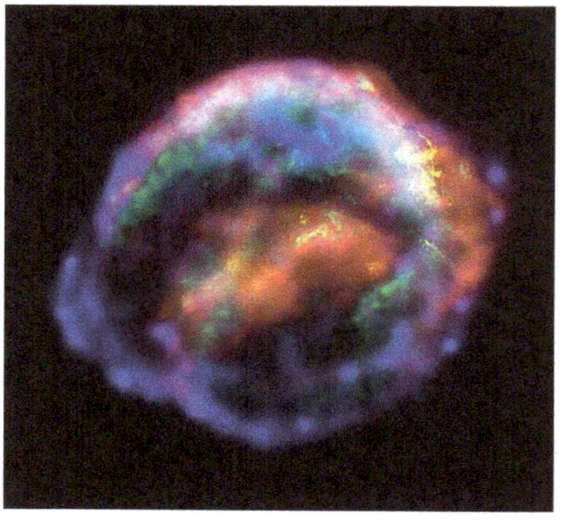

Supernovae resemble novae in several respects. Both are characterized by a tremendous, rapid brightening lasting for a few weeks, followed by a slow dimming. Spectroscopically, they show blue-shifted emission lines, which imply that hot gases are blown outward. But a supernova explosion, unlike a nova outburst, is a cataclysmic event for a star, one that essentially ends its active (i.e., energy-generating) lifetime. When a star "goes supernova," considerable amounts of its matter, equaling the material of several Suns, may be blasted into space with such a burst of energy as to enable the exploding star to outshine its entire home galaxy.

Supernovae explosions release not only tremendous amounts of radio waves and X-rays but also cosmic rays. Some gamma-ray bursts have been associated with supernovae. Supernovae also release many of the heavier elements that make up the components of the solar system, including Earth, into the interstellar medium. Spectral analyses show that abundances of the heavier elements are greater than normal, indicating that these elements do indeed form during the course of the explosion. The shell of a supernova remnant continues to expand until, at a very advanced stage, it dissolves into the interstellar medium.

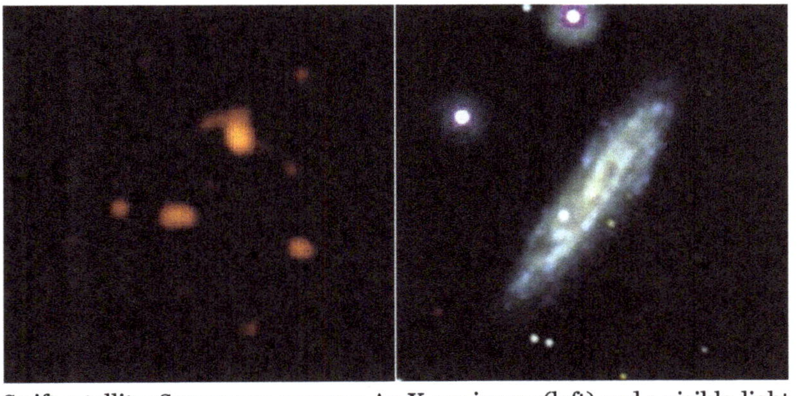

Swift satellite; Supernova 2007uy: An X-ray image (left) and a visible-light image (right) of Supernova 2007uy in galaxy NGC 2770 before Supernova 2008D exploded, image captured by the Swift satellite.

Types of Supernovae

Supernovae may be divided into two broad classes, Type I and Type II, according to the way in which they detonate. Type I supernovae may be up to three times brighter than Type II; they also differ from Type II supernovae in that their spectra contain no hydrogen lines and they expand about twice as rapidly.

Type II Supernovae

The so-called classic explosion, associated with Type II supernovae, has as progenitor a very massive star (a Population I star) of at least eight solar masses that is at the end of its active lifetime. (These are seen only in spiral galaxies, most often near the arms). Until this stage of its evolution, the star has shone by means of the nuclear energy released at and near its core in the process of squeezing and heating lighter elements such as hydrogen or helium into successively heavier elements—i.e., in the process of nuclear fusion. Forming elements heavier than iron absorbs rather than produces energy, however, and, since energy is no longer available, an iron core is built up at the centre of the aging, heavyweight star. When the iron core becomes too massive, its ability to support itself by means of the outward explosive thrust of internal fusion reactions fails to counteract the tremendous pull of its own gravity. Consequently, the core collapses. If the core's mass is less than about three solar masses, the collapse continues until the core reaches a point at which its constituent nuclei and free electrons are crushed together into a hard, rapidly spinning core. This core consists almost entirely of neutrons, which are compressed in a volume only 20 km (12 miles) across but whose combined weight equals that of several Suns. A teaspoonful of this extraordinarily dense material would weigh 50 billion tons on Earth. Such an object is called a neutron star.

The supernova detonation occurs when material falls in from the outer layers of the star and then rebounds off the core, which has stopped collapsing and suddenly presents a hard surface to the infalling gases. The shock wave generated by this collision propagates outward and blows off the star's outer gaseous layers. The amount of material blasted outward depends on the star's original mass.

If the core mass exceeds three solar masses, the core collapse is too great to produce a neutron star; the imploding star is compressed into an even smaller and denser body—namely, a black hole. Infalling material disappears into the black hole, the gravitational field of which is so intense that not even light can escape. The entire star is not taken in by the black hole, since much of the falling envelope of the star either rebounds from the temporary formation of a spinning neutron core or misses passing through the very centre of the core and is spun off instead.

Type I Supernovae

Type I supernovae can be divided into three subgroups—Ia, Ib, and Ic—on the basis of their spectra. The exact nature of the explosion mechanism in Type I generally is still uncertain, although Ia supernovae, at least, are thought to originate in binary systems consisting of a moderately massive star and a white dwarf, with material flowing to the white dwarf from its larger companion. A thermonuclear explosion results if the flow of material is sufficient to raise the mass of the white dwarf above the Chandrasekhar limit of 1.44 solar masses. Unlike the case of an ordinary nova, for which the mass flow is less and only a superficial explosion results, the white dwarf in a Ia supernova explosion is presumably destroyed completely. Radioactive elements, notably nickel-56, are formed. When nickel-56 decays to cobalt-56 and the latter to iron-56, significant amounts of energy are released, providing perhaps most of the light emitted during the weeks following the explosion.

Type Ia supernovae are useful probes of the structure of the universe, since they all have the same luminosity. By measuring the apparent brightness of these objects, one also measures the expansion rate of the universe and that rate's variation with time. Dark energy, a repulsive force that is the dominant component (73 percent) of the universe, was discovered in 1998 with this method. Type Ia supernovae that exploded when the universe was only two-thirds of its present size were fainter and thus farther away than they would be in a universe without dark energy. This implies that the expansion rate of the universe is faster now than it was in the past, a result of the current dominance of dark energy. (Dark energy was negligible in the early universe).

White Dwarfs

White dwarf star is any of a class of faint stars representing the endpoint of the evolution of intermediate- and low-mass stars. White dwarf stars, so called because of the white colour of the first few that were discovered, are characterized by a low luminosity, a mass on the order of that of the Sun, and a radius comparable to that of Earth. Because of their large mass and small dimensions, such stars are dense and compact objects with average densities approaching 1,000,000 times that of water.

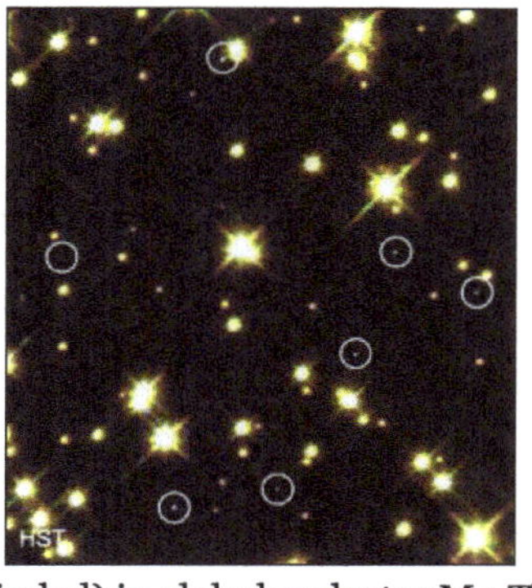

White dwarf stars (circled) in globular cluster M4.The brightest stars in this field are yellow stars similar to the Sun; smaller, dim stars are red dwarfs.

Unlike most other stars that are supported against their own gravitation by normal gas pressure, white dwarf stars are supported by the degeneracy pressure of the electron gas in their interior. Degeneracy pressure is the increased resistance exerted by electrons composing the gas, as a result of stellar contraction. The application of the so-called Fermi-Dirac statistics and of special relativity to the study of the equilibrium structure of white dwarf stars leads to the existence of a mass-radius relationship through which a unique radius is assigned to a white dwarf of a given mass; the larger the mass, the smaller the radius. Furthermore, the existence of a limiting mass is predicted, above which no stable white dwarf star can exist. This limiting mass, known as the Chandrasekhar limit, is on the order of 1.4 solar masses. Both predictions are in excellent agreement with observations of white dwarf stars.

The central region of a typical white dwarf star is composed of a mixture of carbon and oxygen. Surrounding this core is a thin envelope of helium and, in most cases, an even thinner layer of hydrogen. A very few white dwarf stars are surrounded by a thin carbon envelope. Only the outermost stellar layers are accessible to astronomical observations.

White dwarfs evolve from stars with an initial mass of up to three or four solar masses or even possibly higher. After quiescent phases of hydrogen and helium burning in its core—separated by a first red-giant phase—the star becomes a red giant for a second time. Near the end of this second red-giant phase, the star loses its extended envelope in a catastrophic event, leaving behind a dense, hot, and luminous core surrounded by a glowing spherical shell. This is the planetary-nebula phase. During the entire course of its evolution, which typically takes several billion years, the star will lose a major fraction of its original mass through stellar winds in the giant phases and through its ejected envelope. The hot planetary-nebula nucleus left behind has a mass of 0.5–1.0 solar mass and will eventually cool down to become a white dwarf.

White dwarfs have exhausted all their nuclear fuel and so have no residual nuclear energy sources. Their compact structure also prevents further gravitational contraction. The energy radiated away into the interstellar medium is thus provided by the residual thermal energy of the nondegenerate ions composing its core. That energy slowly diffuses outward through the insulating stellar envelope, and the white dwarf slowly cools down. Following the complete exhaustion of this reservoir of thermal energy, a process that takes several additional billion years, the white dwarf stops radiating and has by then reached the final stage of its evolution and becomes a cold and inert stellar remnant. Such an object is sometimes called a black dwarf.

White dwarf stars are occasionally found in binary systems, as is the case for the white dwarf companion to the brightest star in the night sky, Sirius. White dwarf stars also play an essential role in Type Ia supernovae and in the outbursts of novae and of other cataclysmic variable stars.

Neutron Stars

Neutron stars comprise one of the possible evolutionary end-points of high mass stars. Once the core of the star has completely burned to iron, energy production stops and the core rapidly collapses, squeezing electrons and protons together to form neutrons and neutrinos. The neutrinos easily escape the contracting core but the neutrons pack closer together until their density is equivalent to that of an atomic nucleus. At this point, the neutrons occupy the smallest space possible (in a similar fashion to the electrons in a white dwarf) and, if the core is less than about 3 solar masses, they exert a pressure which is capable of supporting a star. For masses larger than this, even the pressure of neutrons cannot support the star against gravity and it collapses into a stellar black hole. A star supported by neutron degeneracy pressure is known as a 'neutron star', which may be seen as a pulsar if its magnetic field is favourably aligned with its spin axis.

Neutrons stars are extreme objects that measure between 10 and 20 km across. They have densities of 10^{17} kg/m^3(the Earth has a density of around 5×103 kg/m^3 and even white dwarfs have densities over a million times less) meaning that a teaspoon of neutron star material would weigh around a billion tonnes. The easiest way to picture this is to imagine squeezing twice the mass of the Sun into an object about the size of a small city. The result is that gravity at the surface of the neutron star is around 10^{11} stronger than what we experience here on Earth, and an object would have to travel at about half the speed of light to escape from the star.

The Crab pulsar and nebula formed in a supernova explosion first noted by Chinese astronomers. This X-ray image shows the pulsar and the nebula which is powered mostly through the loss of rotational energy by the neutron star.

Born in a core-collapse supernova explosion, neutron stars rotate extremely rapidly as a consequence of the conservation of angular momentum, and have incredibly strong magnetic fields due to conservation of magnetic flux. The relatively slowing rotating core of the massive star increases its rotation rate enormously as it collapses to form the much smaller neutron star. This is analogous to the increased spin of an iceskater if she concentrates her mass around her spin axis by bringing her arms close to her body. At the same time, the magnetic field lines of the massive star are pulled closer together as the core collapses. This intensifies the magnetic field of the star to around 10^{12} times that of the Earth.

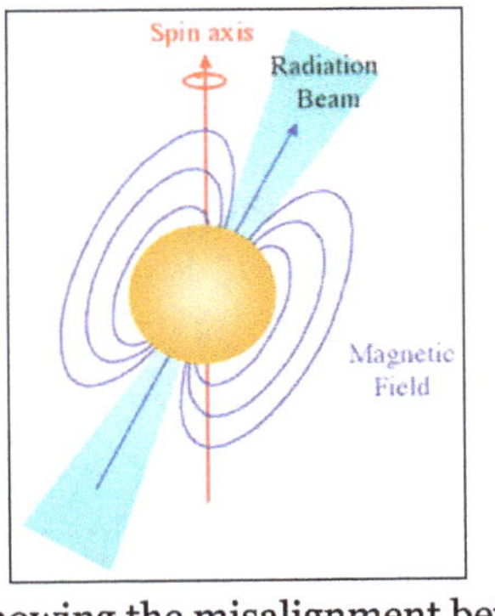

Schematic of a pulsar showing the misalignment between the rotation axis and the radiation beams emitted from the magnetic poles.

The result is that neutron stars can rotate up to at least 60 times per second when born. If they are part of a binary system, they can increase this rotation rate through the accretion of material, to over 600 times per second. Neutron stars that have lost energy through radiative processes have been observed to rotate as slowly as once every 8 seconds while still maintaining radio pulses, and neutron stars that have been braked by winds in X-ray systems can have rotation rates as slow as once every 20 minutes. Observations also reveal that the rotation rate of isolated neutron stars slowly changes over time, generally decreasing as the star ages and rotational energy is lost to the surroundings through the magnetic field (though occasionally glitches are seen). An example is the Crab pulsar, which is slowing its spin at a rate of 38 nanoseconds per day, releasing enough energy to power the Crab nebula.

Astronomers measure these rotation rates by detecting electromagnetic radiation ejected through the poles of the magnetic field. These magnetic poles are generally misaligned with the rotation axis of the neutron star and so the radiation beam sweeps around as the star rotates. This is much the same as the beam of light from a lighthouse sweeping around. If the Earth lies in the path of the beam, we see the neutron star/ pulsar. If not, we see only the supernova remnant. This also nicely accounts for the fact that we do not see a pulsar in every supernova remnant.

Neutron stars do not necessarily exist in isolation, and those that form part of a binary system usually emit strongly in X-rays. X-ray binaries typically result from the transfer of material from a main sequence companion onto the neutron star, while short-duration gamma ray bursts are thought to result from the merger of two neutron stars.

The existence of neutron stars as a result of supernova explosions was tentatively predicted in 1933, one year after the discovery of the neutron as an elementary particle. However, it was not until 1967 that Jocelyn Bell observed the periodic pulses of radio emission characteristic of pulsars. There are now over 1,300 neutron stars known and about 105 predicted to exist in the disk of the Milky Way.

Pulsars

Pulsar or pulsating radio star is any of a class of cosmic objects, the first of which were discovered through their extremely regular pulses of radio waves. Some objects are known to give off short rhythmic bursts of visible light, X-rays, and gamma radiation as well, and others are "radio-quiet" and emit only at X- or gamma-ray wavelengths.

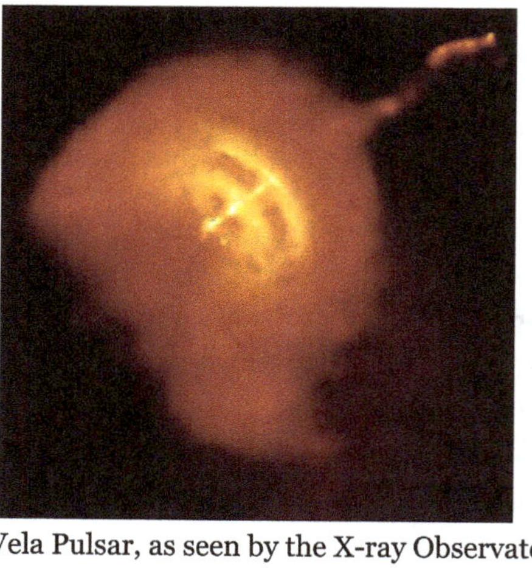

The Vela Pulsar, as seen by the X-ray Observatory.

Characteristics

Pulsars are rapidly spinning neutron stars, extremely dense stars composed almost entirely of neutrons and having a diameter of only 20 km (12 miles) or less. Pulsar masses range between 1.18 and 1.97 times that of the Sun, but most pulsars have a mass 1.35 times that of the Sun. A neutron star is formed when the core of a violently exploding star called a supernova collapses inward and becomes compressed together. Neutrons at the surface of the star decay into protons and electrons. As these charged particles are released from the surface, they enter an intense magnetic field (10^{12} gauss; Earth's magnetic field is 0.5 gauss) that surrounds the star and rotates along with it. Accelerated to speeds approaching that of light, the particles give off electromagnetic radiation by synchrotron emission. This radiation is released as intense beams from the pulsar's magnetic poles.

These magnetic poles do not coincide with the rotational poles, and so the rotation of the pulsar swings the radiation beams around. As the beams sweep regularly past Earth with each complete rotation, an evenly spaced series of pulses is detected by ground-based telescopes. Antony Hewish and Jocelyn Bell, astronomers working at the University of Cambridge, first discovered pulsars in 1967 with the aid of a radio telescope specially designed to record very rapid fluctuations in radio sources. Subsequent searches have resulted in the detection of about 2,000 pulsars. A significant percentage of these objects are concentrated toward the plane of the Milky Way Galaxy, the enormous galactic system in which Earth is located.

Rotation

Although all known pulsars exhibit similar behaviour, they show considerable variation in the length of their periods—i.e., the intervals between successive pulses. The period of the slowest pulsar so far observed is about 11.8 seconds in duration. The pulsar designated PSR J1939+2134 was the fastest known for more than two decades. Discovered in 1982, it has a period of 0.00155 second, or 1.55 milliseconds, which means it is spinning 642 times per second. In 2006 an even faster one was reported: known as J1748−2446ad, it has a period of 1.396 milliseconds, which corresponds to a spin rate of 716 times per second. These spin rates are close to the theoretical limit for a pulsar because a neutron star rotating only about four times faster would fly apart as a result of "centrifugal force" at its equator, notwithstanding a gravitational pull so strong that the star's escape velocity is about half the speed of light.

These fast pulsars are known as millisecond pulsars. They form in supernovae like slower-rotating pulsars. However, millisecond pulsars often occur in binary star systems. After the supernova, the neutron star accretes matter from its companion, causing the pulsar to spin faster.

Period Changes

Careful timing of radio pulsars shows that they are slowing down very gradually at a rate of typically a millionth of a second per year. The ratio of a pulsar's present period to the average slowdown rate gives some indication of its age. This so-called characteristic, or timing, age can be in close agreement with the actual age. For example, the Crab Pulsar, which was formed during a supernova explosion observed in 1054 CE, has a characteristic age of 1,240 years; however, pulsar J0205+6449, which was formed during a supernova in 1181 CE, has a characteristic age of 5,390 years.

Because pulsars slow down so gradually, they are very accurate clocks. Since pulsars also have strong gravitational fields, this accuracy can be used to test theories of gravity. American physicists Joseph Taylor and Russell Hulse won the Nobel Prize for Physics in 1993 for their study of timing variations in the pulsar PSR 1913+16. PSR 1913+16 has a companion neutron star with which it is locked in a tight orbit. The two stars' enormous interacting gravitational fields affect the regularity of the radio pulses, and, by timing these and analyzing their variations, Taylor and Hulse found that the stars were rotating ever faster around each other in an increasingly tight orbit. This orbital decay is presumed to occur because the system is losing energy in the form of gravity waves. This was the first experimental evidence for the existence of the gravitational waves predicted by Albert Einstein in his general theory of relativity.

Pulsars also experience much more drastic period changes, which are called glitches, in which the period suddenly increases and then gradually decreases to its pre-glitch value. Some glitches are caused by "starquakes," or sudden cracks in the rigid iron crust of the star. Others are caused by an interaction between the crust and the more fluid interior. Usually the interior is loosely coupled to the crust, so the crust can slow down relative to the interior. However, sometimes the coupling between the crust and interior becomes stronger, spinning up the pulsar and causing a glitch.

Pulsars in Visible Light, X-rays and Gamma Rays

Some pulsars, such as the Crab and Vela pulsars, are losing rotational energy so precipitously that they also emit radiation of shorter wavelength. The Crab Pulsar appears in optical photographs as a moderately bright (magnitude 16) star in the centre of the Crab Nebula. Soon after the detection of its radio pulses in 1968, astronomers at the Steward Observatory in Arizona found that visible light from the Crab Pulsar flashes at exactly the same rate. The star also produces regular pulses of X-rays and gamma rays. The Vela Pulsar is much fainter at optical wavelengths (average magnitude 24) and was observed in 1977 during a particularly sensitive search with the large Anglo-Australian Telescope situated at Parkes, Australia. It also pulses at X-ray wavelengths. The Vela Pulsar does, however, give off gamma rays in regular pulses and is the most intense source of such radiation in the sky.

Some X-ray pulsars are "accreting" pulsars. These pulsars are in binaries, and the neutron star accretes material from its companion. This material flows to the magnetic polar caps, where it releases X-rays. Another class of X-ray pulsars is called "anomalous." These pulsars have periods of more than five seconds, sometimes give off bursts of X-rays, and are often associated with supernova remnants. These pulsars arise from highly magnetized neutron stars, or magnetars, which have a magnetic field of between 10^{14} and 10^{15} gauss. (The magnetars also have been identified with another class of objects, the soft gamma-ray repeaters, which give off bursts of gamma rays).

Some pulsars emit only in gamma rays. In 2008 the Fermi Gamma-ray Space Telescope discovered the first such pulsar within the supernova remnant CTA 1; since then it has found 11 others. Unlike radio pulsars, the gamma-ray emission does not come from the particle beams at the poles but arises far from the neutron star surface. The precise physical process that generates the gamma-ray pulses is unknown.

References

- Star-astronomy, science: britannica.com, Retrieved 06 January, 2019
- What-is-the-chemical-composition-of-most-stars-12731968: sciencing.com, Retrieved 15 May, 2019
- Stellar-magnetic-fields, businesses-and-occupations, economics-business-and-labor, social-sciences-and-law: encyclopedia.com, Retrieved 27 June, 2019
- Stellar-nucleosynthesis-2699311: thoughtco.com, Retrieved 04 February, 2019
- White-dwarf-star, science: britannica.com, Retrieved 18 April, 2019

3

Solar Physics

The field of science that specializes in the study of the sun is referred to as solar physics. Solar cycle, solar-terrestrial effects, solar magnetic fields, solar wind, solar flares, magnetic flux emergence, etc. fall under its domain. The topics elaborated in this chapter will help in gaining a better perspective about solar physics.

Solar physics is the study of the fundamental processes occurring in the sun. Primarily this is related to the dynamics of plasmas and their interplay with the sun's magnetic fields, and how these processes vary in different regions of the sun, from its core to the surrounding corona. Sun is considered as a special star since it is the one which provides most of the energy to earth. It is also the only star which can resolve the spatial scale.

Many of the properties of sun is still a mystery and study of it is thus of great importance. A close track on the sun is really important since atmosphere and activities of sun have a great deal of impact on earth.

The study of various properties of sun ranges back from ancient times. The ancient calendars were based on solar eclipse and sunspots. Ancient people also used solar observations for timekeeping and for navigation purposes. It was based on solar observations that Nicolas Copernicus put forward heliocentric model. In modern day research on sun is to study the sunspot, coronal heat problem etc.

Various space agencies like NASA (National Aeronautics and space administration) and ESA (European Space Agency) have launched many satellites to study various properties of sun. In February 2010 NASA launched Solar Dynamic Observatory in order to study activities on sun.

In December 1995 both the space agencies (NASA and ESA) jointly launched Solar and Heliosphere Observatory in order to study about the solar wind. In order to find how corona and magnetic field of sun interact with each other Japanese space agency launched HINODE in the year 2006.

Solar physics is related to many other branches of physics like atomic and molecular physics, astrophysics etc. The sun acts as a laboratory for the study of plasma.

The Sun

Sun is a star around which Earth and the other components of the solar system revolve.

It is the dominant body of the system, constituting more than 99 percent of its entire mass. The Sun is the source of an enormous amount of energy, a portion of which provides Earth with the light and heat necessary to support life.

The Sun is classified as a G2 V star, with G2 standing for the second hottest stars of the yellow G class—of surface temperature about 5,800 kelvins (K)—and the V representing a main sequence, or dwarf, star, the typical star for this temperature class. (G stars are so called because of the prominence of a band of atomic and molecular spectral lines that the German physicist Joseph von Fraunhofer designated G). The Sun exists in the outer part of the Milky Way Galaxy and was formed from material that had been processed inside a supernova. The Sun is not, as is often said, a small star. Although it falls midway between the biggest and smallest stars of its type, there are so many dwarf stars that the Sun falls in the top 5 percent of stars in the neighbourhood that immediately surrounds it.

Physical Properties

The radius of the Sun, $R_{\odot}$, is 109 times that of Earth, but its distance from Earth is 215 $R_{\odot}$, so it subtends an angle of only 1/2° in the sky, roughly the same as that of the Moon. By comparison, Proxima Centauri, the next closest star to Earth, is 250,000 times farther away, and its relative apparent brightness is reduced by the square of that ratio, or 62 billion times. The temperature of the Sun's surface is so high that no solid or liquid can exist there; the constituent materials are predominantly gaseous atoms, with a very small number of molecules. As a result, there is no fixed surface. The surface viewed from Earth, called the photosphere, is the layer from which most of the radiation reaches us; the radiation from below is absorbed and reradiated, and the emission from overlying layers drops sharply, by about a factor of six every 200 kilometres (124 miles). The Sun is so far from Earth that this slightly fuzzy surface cannot be resolved, and so the limb (the visible edge) appears sharp.

The eight planets of the solar system and Pluto, in a montage of images scaled to show the approximate sizes of the bodies relative to one another. Outward from the Sun, which is represented to scale by the yellow segment at the extreme left, are the four rocky terrestrial planets (Mercury, Venus, Earth, and Mars), the four hydrogen-rich giant planets (Jupiter, Saturn, Uranus, and Neptune), and icy, comparatively tiny Pluto.

The mass of the Sun, $M_\odot$, is 743 times the total mass of all the planets in the solar system and 330,000 times that of Earth. All the interesting planetary and interplanetary gravitational phenomena are negligible effects in comparison to the force exerted by the Sun. Under the force of gravity, the great mass of the Sun presses inward, and to keep the star from collapsing, the central pressure outward must be great enough to support its weight. The density at the Sun's core is about 100 times that of water (roughly six times that at the centre of Earth), but the temperature is at least 15,000,000 K, so the central pressure is at least 10,000 times greater than that at the centre of Earth, which is 3,500 kilobars. The nuclei of atoms are completely stripped of their electrons, and at this high temperature they collide to produce the nuclear reactions that are responsible for generating the energy vital to life on Earth.

While the temperature of the Sun drops from 15,000,000 K at the centre to 5,800 K at the photosphere, a surprising reversal occurs above that point; the temperature drops to a minimum of 4,000 K, then begins to rise in the chromosphere, a layer about 7,000 kilometres high at a temperature of 8,000 K. During a total eclipse the chromosphere appears as a pink ring. Above the chromosphere is a dim, extended halo called the corona, which has a temperature of 1,000,000 K and reaches far past the planets. Beyond a distance of $5R_\odot$ from the Sun, the corona flows outward at a speed (near Earth) of 400 kilometres per second (km/s); this flow of charged particles is called the solar wind.

The Sun is a very stable source of energy; its radiative output, called the solar constant, is 1.366 kilowatts per square metre at Earth and varies by no more than 0.1 percent. Superposed on this stable star, however, is an interesting 11-year cycle of magnetic activity manifested by regions of transient strong magnetic fields called sunspots.

Structure of the Sun

Astrophysicists classify the Sun as a star of average size, temperature, and brightness—a typical dwarf star just past middle age. It has a power output of about 10^{26} watts and is expected to continue producing energy at that rate for another 5 billion years. The Sun is said to have a diameter of 1.4 million kilometers, about 109 times the diameter of Earth, but this is a slightly misleading statement because the Sun has no true "surface." There is nothing hard, or definite, about the solar disk that we see; in fact, the matter that makes up the apparent surface is so rarified that we would consider it to be a vacuum here on Earth. It is more accurate to think of the Sun's boundary as extending far out into the solar system, well beyond Earth. In studying the structure of the Sun, solar physicists divide it into four domains: the interior, the surface atmospheres, the inner corona, and the outer corona.

The Interior

The Sun's interior domain includes the core, the radiative layer, and the convective layer. The core is the source of the Sun's energy, the site of thermonuclear fusion. At a temperature of about 15,000,000 K, matter is in the state known as plasma: atomic nuclei (principally protons) and electrons moving at very high speeds. Under these conditions two protons can collide, overcome their electrical repulsion, and become cemented together by the strong nuclear force. This process is known as nuclear fusion, and it results in the formation of heavier elements as well as the release of energy in the form of gamma ray photons. The energy output of the Sun's core is so large that it would shine about 1013 times brighter than the solar surface if we could "see" it.

The immense energy produced in the core is bound by the surrounding radiative layer. This layer has an insulating effect that helps maintain the high temperature of the core. The gamma photons produced by fusion in the core are absorbed and re-emitted repeatedly by nuclei in the radiative layer, with the re-emitted photons having successively lower energies and longer wavelengths. By the time the photons leave the Sun, their wavelengths are mostly in the visible range. The energy produced in the core can take as long as 50 million years to work its way through the radiative layer of the Sun. If the processes in the core of the Sun suddenly stopped, the surface would continue to shine for millions of years.

Above the radiative layer is the convective layer where the temperature is lower, and radiation is less significant. Energy is transported outward mostly by convection. Hot regions at the bottom of this layer become buoyant and rise. At the same time, cooler material from above descends, and giant convective cells are formed. This convection is widespread throughout the Sun, except in the core and radiative layer where the temperature is too high. The tops of convective cells can be seen on the photosphere as granules. Convective circulation of plasma (charged particles) generates large magnetic fields that play an important role in producing sunspots and flares.

Thermonuclear Fusion

The nuclear fusion, now occurring in the core of the Sun, turns hydrogen nuclei into helium nuclei. In fact, that is how the elements heavier than hydrogen are made; the thermonuclear fusion at the core of stars can produce the first 26 elements, up to iron. The Sun, because of its relatively small mass, will go through only the first two stages of fusion, the hydrogen-helium stage and the helium-carbon stage.

Hydrogen-helium fusion can occur in more than one way, but in any case the temperature must be in the vicinity of 15 million K so that two positively charged particles will be moving fast enough to overcome their electrical repulsion when they collide. The density must be large, and the immense solar gravity compresses the gas so that it is ten times as dense as gold at the center of the Sun. If the two particles can get close enough together, the very short-range strong nuclear force will take effect and fuse them together. The most common fusion reaction in the Sun is shown in figure.

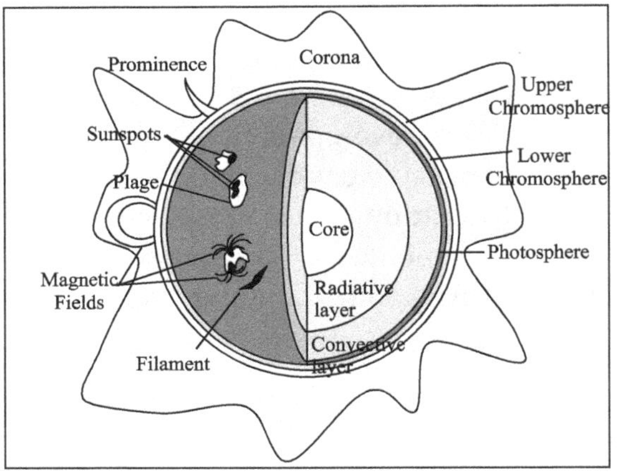

The structure of the Sun.

If we compare the total mass that went into this three-step fusion reaction to the total mass at the end, we will see that a small amount of mass has disappeared. For this reaction, 0.7 percent of the mass disappears and is converted into energy according to $E = mc^2$ (where E = energy, m = mass and c = the speed of light). The actual energy produced from this reaction (for a given 4 Hydrogen atoms) can be found by,

$$E = (0.007)(\text{mass of } 4H)c^2.$$

In order to produce the known energy output of the Sun, 700 million tons of hydrogen are fused into 695 million tons of helium each second! It may be shocking to think that the Sun is losing mass at the rate of 5 million tons per second, but its total mass is so great that this rate of loss can continue for a long time.

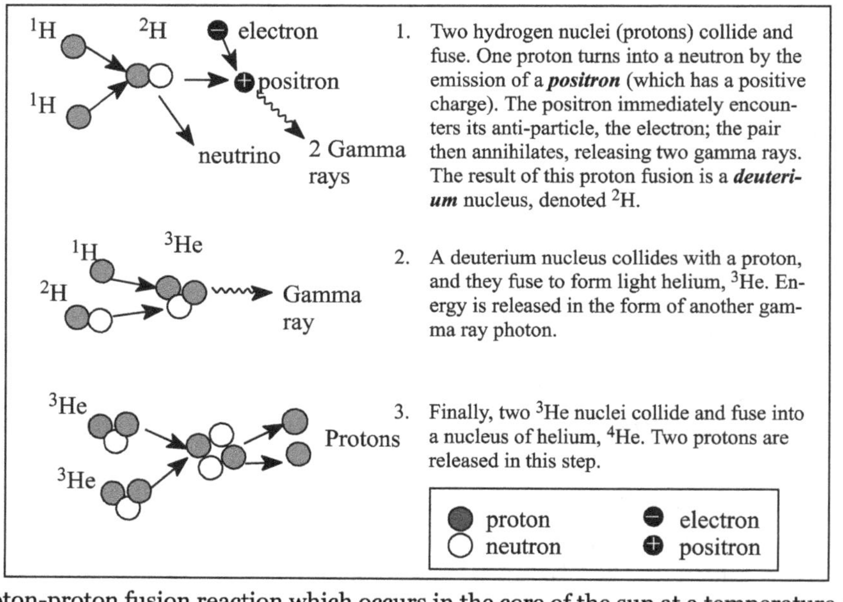

The proton-proton fusion reaction which occurs in the core of the sun at a temperature of about 15,000,000 K. In this reaction 0.7% of the total mass disappears and is released as energy.

Scientists have dreamed of being able to harness fusion energy to produce electricity on Earth. In attempting the fusion process we are trying to duplicate the conditions in the interior of a star. There are significant problems associated with handling plasma at 10 to 15 million degrees. The only "container" that can hold material at such high temperatures is a magnetic container. At present, fusion experiments involve the confinement of plasma in very large toroidal (donut-shaped) magnetic fields produced by devices called tokamaks. These devices have produced small scale fusion, but the energy input still far outweighs the energy output. The most promising fusion reactions are the deuterium-deuterium reaction (D-D) and deuterium-tritium reaction (D-T). Unlike the fusion process in the Sun, we do not attempt the first step in which two protons fuse to form deuterium. This collision has a very low cross-section, meaning that it is very unlikely. The deuterium fuel for Earth-based fusion is extracted from water, which contains a small percent of deuterium and tritium. The D-T reaction has a higher cross-section, making it easier to achieve, but it produces extra neutrons, which makes it more dangerous.

It should be understood that we have achieved uncontrolled fusion here on Earth in the form of the hydrogen bomb. Early nuclear weapons, like those used at Hiroshima and Nagasaki in 1945, were nuclear fission devices which used ^{235}U as an energy source. Today these fission bombs, sometimes incorrectly called atomic bombs, are used to trigger the larger fusion reaction which turns hydrogen into helium and produces a large amount of energy in one short burst. At the site of such a detonation, the conditions resemble the core of a star with temperatures reaching about 15 million degrees.

The Surface Atmospheres

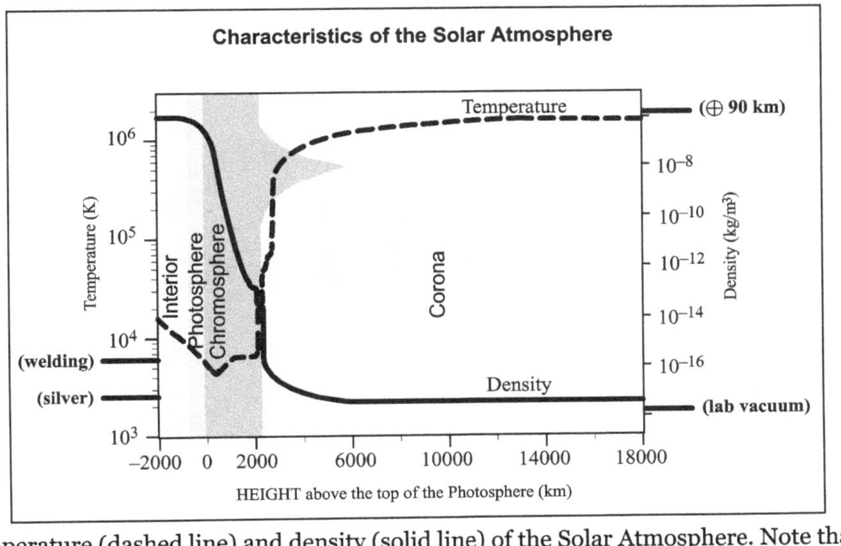

Temperature (dashed line) and density (solid line) of the Solar Atmosphere. Note that the highest density on the scale here is still only as dense as the Earth's atmosphere at 90 km up. The melting temperature of silver is near the bottom of the temperature scale shown here.

The solar surface atmospheres are composed of the photosphere and the chromosphere. The photosphere is the part of the Sun that we see with our eyes—it produces most of the visible (white) light. Bubbles of hotter material well up from within the Sun, dividing the surface of the photosphere into bright granules that expand and fade in several minutes, only to be replaced by the next upwelling. The photosphere is one of the coolest layers of the Sun; its temperature is about 6,000 K.

Sometimes huge magnetic-field bundles break through the photosphere, disturbing this boiling layer with a set of conditions known collectively as solar activity. These magnetic fields create cooler, darker regions, which we see as sunspots. Early observers of sunspots quickly noted that they appear to migrate across the disk of the Sun as it rotates.

The Sun's rotation rate differs according to latitude: as seen from the Earth, the equatorial region rotates with a period of about 27 days, while the rotational period closer to the poles is about 32 days.

The Sun's Vital Statistics

- Age: At least 4.5 billion years in present state.
- Chemical composition of photosphere (by mass, in percent):
    - Hydrogen - 73.46.
    - Helium - 24.85.
    - Oxygen - 0.77.
    - Carbon - 0.29.
    - Iron - 0.16.
    - Neon - 0.12.
    - Nitrogen - 0.09.
    - Silicon - 0.07.
    - Magnesium - 0.05.
    - Sulfur - 0.04.
    - Other - 0.10.
- Density (water=1000):
    - Mean density of entire Sun - 1410 kg/m^3.
    - Interior (center of Sun) - 160000 kg/m^3.
    - Surface (photosphere) - 10–6 kg/m^3.

- Chromosphere - 10–9 kg/m^3.
- Low corona - 10–13 kg/m^3.
- Sea level atmosphere of Earth (for comparison) - 1.2 kg/m^3.

- Diameter (measured at the Photosphere): $1.39 x 10^6$ km (or 109 times the diameter of Earth and 9.75 times the diameter of Jupiter, the largest planet.
- Distance:
  - Mean distance from Earth - 1.5 x 10^8 km.
  - Variation in distance through the year - ±1.5 percent.
- Magnetic field strengths for typical features:
  - Sunspots - 0.3 tesla.
  - Polar field - 10^{-4} tesla.
  - Bright, chromospheric network - 0.0025 tesla.
  - Ephemeris (unipolar) active regions - 0.0020 tesla.
  - Chromospheric plages - 0.02 tesla.
  - Prominences - 10^{-3} to 10^{-2} tesla.
  - Earth (for comparison) - 7 x 10^{-5} tesla at pole.
- Mass: $1.99 x 10^{30}$ kg (or 333 000 times the mass of Earth).
- Rotation (as seen from Earth):
  - Of solar equator - 26.8 days.
  - At solar latitude 30 - 28.2 days.
  - At solar latitude 60 - 30.8 days.
  - At solar latitude 75 - 31.8 days.
- Solar radiation:
  - Entire Sun - 3.83 x 10^{23} kW.
  - Unit area of surface of Sun - 6.29 x 104 kW/m^2.
  - Received at top of Earth's atmosphere - 1370 W/m^2.
- Surface brightness of the Sun (photosphere):
  - Compared to full Moon - 398 000 times.

    - Compared to inner corona - 300 000 times.
    - Compared to outer corona - 10^{10} times.
    - Compared to daytime sky on Pikes Peak - 100 000 times.
    - Compared to daytime sky at Orange, N.J. - 1000 times.
- Temperature:
    - Interior (center) - 15 000 000 K.
    - Surface (photosphere) - 6050 K.
    - Sunspot umbra (typical) - 4240 K.
    - Penumbra (typical) - 5680 K.
    - Chromosphere - 4300 to 50 000 K.
    - Corona - 800 000 to 3 000 000 K.
- Volume: 1.41×10^{27} m^3 (or 1.3 million times the volume of Earth).

These periods can easily be determined by watching sunspots over several days. It is now known, however, that these periods correspond to the photosphere where the sunspots reside, and that the rotational period varies in the different layers above the photosphere. This complicated variation of rotational period according to latitude and depth contributes to the shearing and twisting that give rise to solar activity.

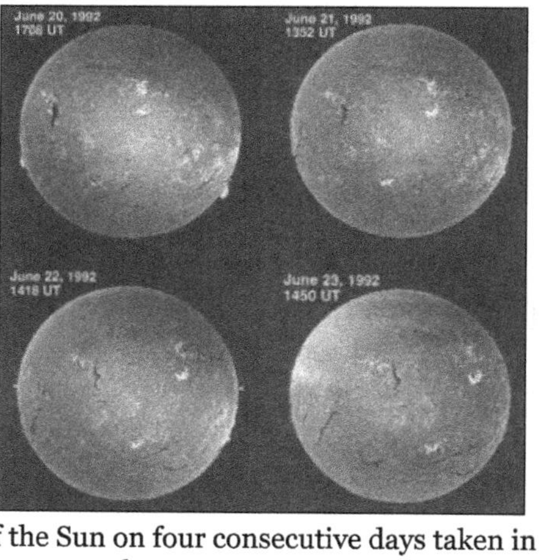

Photos of the Sun on four consecutive days taken in H light.
Features can be seen to move as the Sun rotates.

The chromosphere lies just above the photosphere, and is slightly cooler at its base. It is called chromo because of its color, which can only be seen when the much brighter light from the photosphere is eliminated. When a solar eclipse occurs, the red chromosphere is seen briefly just before and after the period of total eclipse. When viewed in white light, the chromosphere is transparent to the brilliant light emitted underneath it by the photosphere. But when viewed only in the red light produced by hydrogen (called H), the chromosphere is seen to be alive with many distinctive features, including long dark filaments and bright areas known as plage that surround sunspot regions. The chromosphere is also characterized by cellular convection patterns, but these cells are much larger than the granules of the photosphere. Near the boundaries of these cells are concentrated magnetic fields that produce vertical jets of material called spicules. Although spicules are considered to be small features of the quiet sun, they are actually about the size of Earth! Flares are much larger and more explosive. The active regions associated with sunspots produce strong magnetic fields, which arch up through the chromosphere and become conduits for material when explosive flares erupt. The cause and timing of these eruptions are of great interest to scientists but are not well understood.

Solar activity is very apparent in the chromosphere, and has a wide range of time scales. Flares begin in seconds and end after minutes or hours. Active regions last many weeks, and may flare many times before fading away. The number of sunspots and active regions rises and falls in a mysterious 11-year cycle. Behind all of these phenomena and time scales are the Sun's magnetic fields, deriving their energy from the interplay of the Sun's rotational and convective motions. The magnetic fields are always changing, yet there is a 22-year magnetic cycle that seems to underlie all of the Sun's activity. The activity that we can observe on the photosphere and chromosphere is merely a "symptom" of what is happening inside the Sun. Although we have many clues, the detailed physics of stellar interiors is still largely a mystery.

The Inner Corona

The inner corona is the wispy halo, extending more than a million kilometers out into space that can be seen when the brilliant disk of the Sun is blocked by the Moon during a total eclipse. The cause of the high temperature of the corona, about 2,000,000 K, is not well understood. The corona is a large source of x-rays which do not penetrate Earth's atmosphere. With instruments on satellites we can look at the corona in x-ray wavelengths and see many details that do not appear in visible light. From this vantage point it is clear that magnetic arches dominate the structure of the corona. Large and small magnetic active regions glow brightly at x-ray wavelengths, while open magnetic field structures appear as gaping coronal holes. The coronal material is generally confined by closed magnetic field structures, anchored at both ends, but the open field structure of coronal holes allows the corona to escape freely to form fast, low density streams in the solar wind. This material travels outward and causes disturbances in Earth's magnetic field. Because of their effects on Earth, we would like to be able to predict when and where coronal holes will form, but as yet we cannot do this.

The Outer Corona

The outer corona extends to Earth and beyond. Its existence is not immediately obvious, since it cannot be seen directly; astrophysicists did not become aware of it until the 1950's. Watching the behavior of comets, Ludwig Biermann realized in the early 1950's that the solar corona must be expanding outward. By 1958, Eugene Parker concluded from theoretical models that particles streaming off the Sun were necessary to maintain the dynamic equilibrium of the corona. Parker's mathematical prediction that particles streamed from the Sun at speeds of several hundred kilometers per second was verified in the early 1960s when satellites detected coronal outflow. This outflow came to be called the solar wind and its speed was accurately measured in 1962 by the Mariner 2 spacecraft bound for Venus. As Parker had predicted, this speed averaged about 400 km/s.

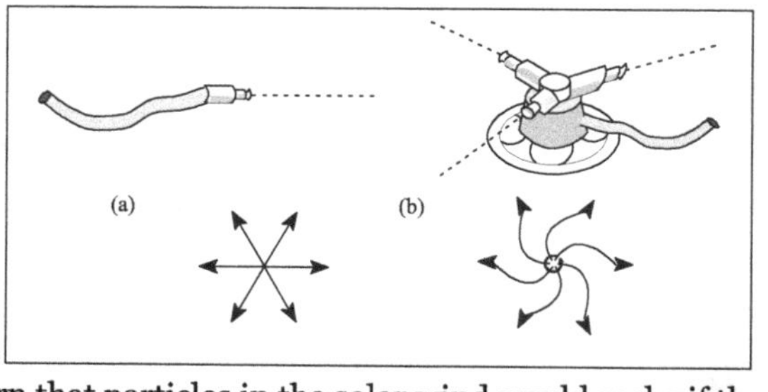

(a) The pattern that particles in the solar wind would make if the Sun did not rotate; the outflowing wind is radial. An analogy would be a garden hose. (b) The "Archimedes spiral" pattern produced by solar rotation.

An analogy would be a garden sprinkler. During the four days that it takes the solar wind to travel to Earth, the Sun rotates about 60 degrees.

In the 30 years since the discovery of the solar wind, we have learned much more about it, and its effects on Earth. The solar wind streams radially outward from the Sun. Solar rotation swings the source around so that the individual streams describe Archimedian spirals; the solar wind speed and density vary according to the conditions on the Sun. This variation in the solar wind intensity began to make more sense after the discovery of coronal holes during the Skylab missions in the early 1970s. Using an x-ray telescope, the Skylab astronauts took many pictures of the Sun which showed coronal holes as large, dark regions with open magnetic field lines where the corona streams outward. These regions grow and shrink, and move around on the Sun in ways that are not yet understood. When a coronal hole is facing Earth, the solar wind reaching Earth is more intense.

The nature of the solar wind is also determined by flare and prominence activity on the Sun. During times of high activity, plasma is hurled off the Sun in vast eruptions that are energized by the turbulent magnetic fields in the inner corona. If ejected mass travels outward and strikes the Earth, we can feel many effects.

The Photosphere

Although there are no fires on the surface of the Sun, the photosphere seethes and roils, displaying the effects of the underlying convection. Photons flowing from below, trapped by the underlying layers, finally escape. This produces a dramatic drop in temperature and density. The temperature at the visible surface is about 5,800 K but drops to a minimum about 4,000 K at approximately 500 kilometres above the photosphere. The density, about 10^{-7} gram per cubic centimetre (g/cm^3), drops a factor of 2.7 every 150 kilometres. The solar atmosphere is actually a vacuum by most standards; the total density above any square centimetre is about 1 gram, about 1,000 times less than the comparable mass in the atmosphere of Earth. One can see through the atmosphere of Earth but not through that of the Sun because the former is shallow, and the molecules absorb only radiation that lies outside of the visible spectrum. The hot photosphere of the Sun, by contrast, contains an ion called negative hydrogen, H^-, a hydrogen nucleus with two electrons attached. The H^- ion absorbs radiation voraciously through most of the spectrum.

The photosphere is the portion of the Sun seen in ordinary light. Its image reveals two dominant features, a darkening toward the outermost regions, called limb darkening, and a fine rice-grain-like structure called granulation. The darkening occurs simply because the temperature is falling; when one looks at the edge of the Sun, one sees light from higher, cooler, and darker layers. The granules are convective cells that bring energy up from below. Each cell measures about 1,500 kilometres across. Granules have a lifetime of about 25 minutes, during which hot gas rises within them at speeds of about 300 metres per second. They then break up, either by fading out or by exploding into an expanding ring of granules. The granules occur all over the Sun. It is believed that the explosion pattern shapes the surrounding granules in a pattern called mesogranulation, although the existence of that pattern is in dispute. A larger, undisputed pattern called supergranulation is a network of outward velocity flows, each about 30,000 kilometres across, which is probably tied to the big convective zone rather than to the relatively small granules. The flow concentrates the surface magnetic fields to the supergranulation-cell boundaries, creating a network of magnetic-field elements.

The photospheric magnetic fields extend up into the atmosphere, where the supergranular pattern dominates the conducting gas. While the temperature above the average surface areas continues to drop, it does not fall as rapidly as at the network edges, and a picture of the Sun at a wavelength absorbed somewhat above the surface shows the network edges to be bright. This occurs throughout the ultraviolet.

Fraunhofer was the first to observe the solar spectrum, finding emission in all colours with many dark lines at certain wavelengths. He assigned letters to these lines, by which some are still known, such as the D-lines of sodium, the G-band, and the K-lines of ionized calcium. But it was the German physicist Gustav R. Kirchhoff who explained the meaning of the lines, explaining that the dark lines formed in cooler upper layers, absorbing the light emerging from below. By comparing these lines with laboratory data, we can identify the elements responsible and their state of ionization and excitation.

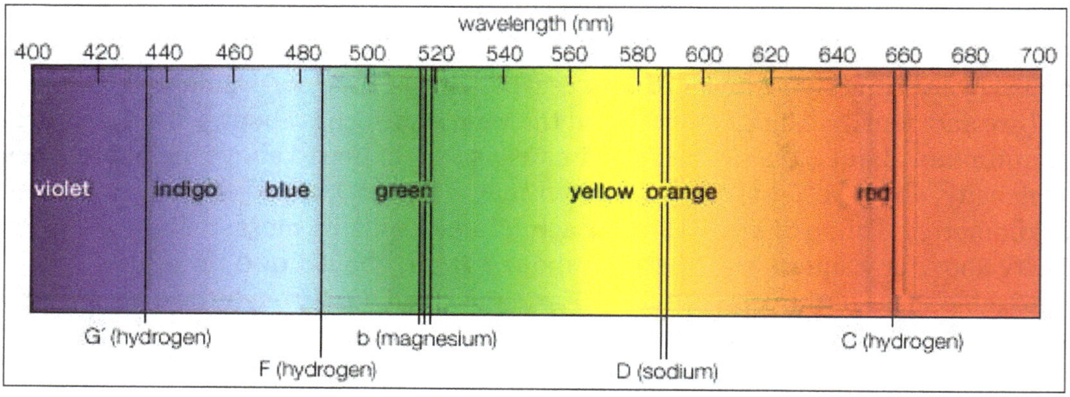

The visible solar spectrum, with prominent Fraunhofer lines representing wavelengths at which light is absorbed by elements.

The spectral lines seen are those expected to be common at 6,000 K, where the thermal energy of each particle is about 0.5 volt. The most abundant elements, hydrogen and helium, are difficult to excite, while atoms such as iron, sodium, and calcium have many lines easily excited at this temperature. When Cecilia Payne, a British-born graduate student studying at Harvard College Observatory in Cambridge, Massachusetts, U.S., recognized the great abundance of hydrogen and helium in 1925, she was persuaded by her elders to mark the result as spurious; only later was the truth recognized. The strongest lines in the visible spectrum are the H^- and K^- (Fraunhofer's letters) lines of ionized calcium. This happens because calcium is easily ionized, and these lines represent transitions in which energy is absorbed by ions in the ground, or lowest energy, state. In the relatively low density of the photosphere and higher up, where atoms are only illuminated from below, the electrons tend to fall to the ground state, since excitation is low. The sodium D-lines are weaker than Ca K because most of the sodium is ionized and does not absorb radiation.

The intensity of the lines is determined by both the abundance of the particular element and its state of ionization, as well as by the excitation of the atomic energy level involved in the line. By working backward one can obtain the abundance of most of the elements in the Sun. This set of abundances occurs with great regularity throughout the universe; it is found in such diverse objects as quasars, meteorites, and new stars. The Sun is roughly 90 percent hydrogen by number of atoms and 9.9 percent helium. The remaining atoms consist of heavier elements, especially carbon, nitrogen, oxygen, magnesium, silicon, and iron, making up only 0.1 percent by number.

Solar Granulation

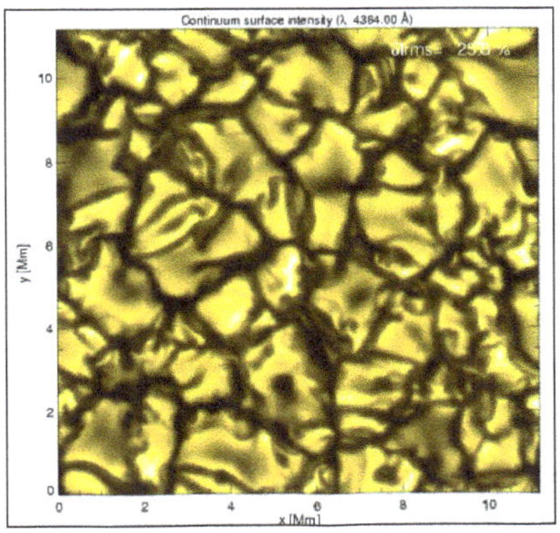

The solar granulation is the process of formation of granules on the photosphere of the sun due to the thermal currents in the thermal columns. The granules will appear as grainy structures.

The thermal currents will go through the plasma of thermal columns. For this process to happen, the spontaneous formation of the plume should happen. This triggers the formation of the granules. Hence the upward motion of gas occurs to fix the mass continuity. This will result dark centers and bright edges.

The solar granules are highly turbulent due to the higher Reynolds number. Where in the intergranular turbulence will be less. The turbulence in divergent flows gets suppressed. Even in the granular as well as inter granular boarders should have turbulence to form solar granulation.

The solar granules will explode due to the granular up flow. This explosion of granules starts at the upper layer of the sun. During the explosion, the higher region becomes bright and the center will become dark and this will get added to the continuum layer. This process is known as the inverted granulation. While the explosion the flow direction will also get inverted. While in explosion not all the dark matter increase in size, instead, they move out of the granules.

These dark dots indicate the downward motion of plumes. The granule explosion is closely related to the super granulation. For the stabilization of explosion, the plumes will rotate with the help of Coriolis force. The dark dots which cross the granules are stable quantities.

Faculae

Facula, in astronomy, is a bright granular structure on the Sun's surface that is slightly hotter or cooler than the surrounding photosphere. A sunspot always has an associated facula, though faculae may exist apart from such spots. Faculae are visible in ordinary white light near the Sun's limb (apparent edge), where the photospheric background is dimmer than near the centre of the disk. The extensions of faculae up into the chromosphere become visible over the entire disk in spectroheliograms taken at the wavelengths of hydrogen or ionized calcium vapour. When seen thus away from the limb, they are called chromospheric faculae or plages.

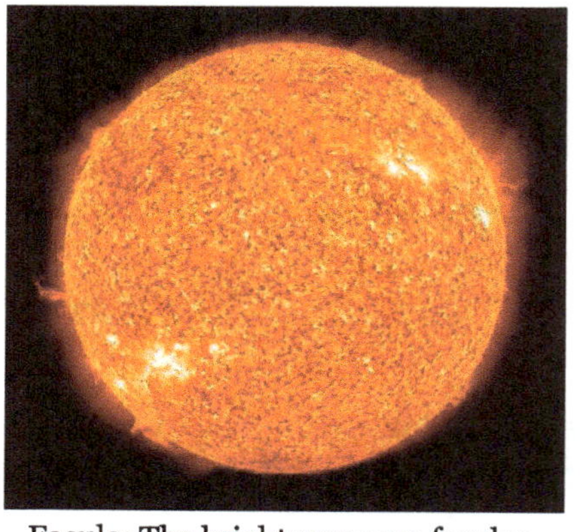

Facula: The bright areas are faculae.

Sunspots

Sunspots form on the surface of the Sun due to strong magnetic field lines coming up from within the Sun trough the solar surface and appear visibly as dark spots compared to their surroundings. These sunspots which can become many times bigger than the Earth are always dark because they are much cooler than the surrounding surface of the Sun itself. A big sunspot can have a temperature of 3700 °C. This sounds like much but if we compare this with the temperature of the photosphere of the Sun which is about 5500 °C, then you see that there is a considerable difference. As a matter of fact, if we could take a sunspot out of the Sun and place it into our night sky it would only be as bright as the full moon, a very big contrast with the bright Sun itself.

Sunspots are a common sight on our Sun during the years around solar maximum. Solar maximum or solar max is the period of greatest solar activity in the solar cycle of the Sun, where one solar cycle lasts about 11 years. Around solar minimum, only very few or even no sunspots can be found. Sunspots form where magnetic field lines come up from the Sun's interior trough the solar surface meaning that every sunspot has its own polarity.

A sunspot consists of two parts:

- The dark part (umbra).
- Lighter part around the dark part (penumbra).

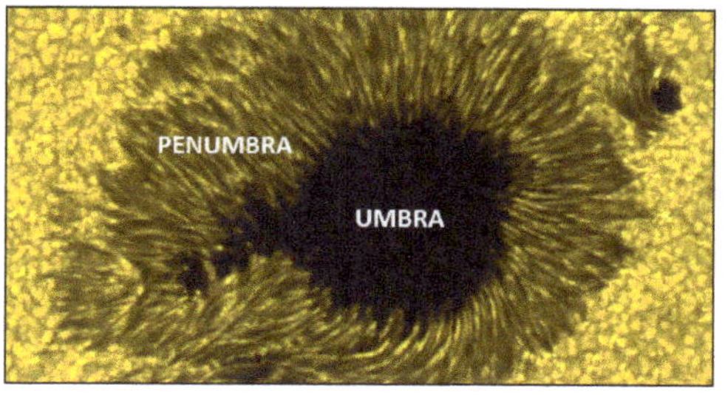

Sunspot Regions

The image below shows a large and complex group with numerous sunspots. This is called a sunspot group or an Active Region. Every day, all the sunspot regions on the Earth-facing solar disk are analysed for their eruptive threat and receive a number. This is done by the NOAA Space Weather Prediction Center. Sunspot regions, especially those with complex magnetic layouts, are known to cause a phenomenon called solar flares.

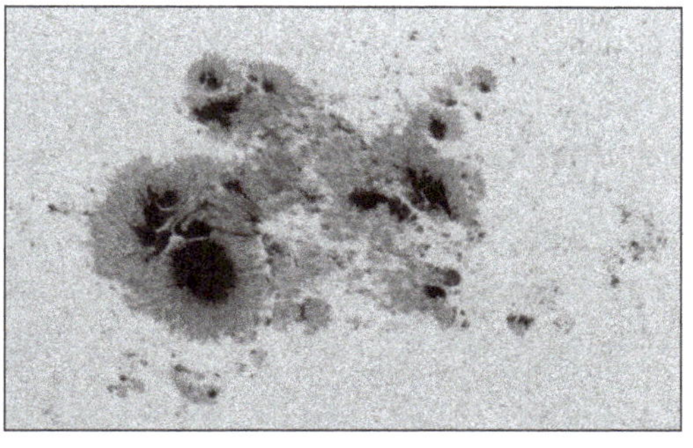

Rotation of the Sun

The Sun rotates around its axis just like Earth does. Solar features on the Sun like sunspot regions follow the rotation of the Sun. This means that a sunspot region travels across the solar disk from east to west as seen from Earth. This is important because sunspot regions need to be close to the central meridian (as seen from Earth) in order to be able to send coronal mass ejection towards Earth. It takes a sunspot region near the equator about 2 weeks to move from the east limb to west limb as seen from Earth. The further away a sunspot region is from the equator the longer it takes move across the face of the Sun. This is because the Sun rotates faster at its equator than at its poles. The rotational period is approximately 25.6 days at the equator and 33.5 days at the poles. Viewed from Earth as it orbits the Sun, the apparent rotational period of the Sun at its equator is about 28 days.

Prominences

The Sun's fiery hot sphere produces a variety of special features. Maybe the most intriguing is a so-called solar prominence. Astronomy enthusiasts, eclipse chasers, and Sun watchers talk about solar prominences all the time.

Think of the Sun as a highly magnetic ball of gas. Following along magnetic field lines, giant loops of plasma can form that extend far outward into space from the Sun's glowing "surface," its photosphere. These loops of plasma are solar prominences, and they can measure many times larger than Earth's diameter.

Prominences push out into the Sun's thin, extremely hot atmosphere, called the corona. They radiate at lower temperatures than the corona itself. These prominences can last for days, sometimes even months, and are often associated with a very powerful surge of gas that flows outward in a giant solar "belch" — a coronal mass ejection that can light up Earth's skies with an aurora and even interfere with our technology.

For solar observers, prominences are a joy. Of course to ever look at the Sun in any way, you need to do it properly and carefully. A solar filter that permits seeing prominences is a so-called Hydrogen-alpha (H-alpha) filter, which transmits the wavelength of light prominences give off.

If you have a telescope fitted with an H-alpha filter, you can see these tongues of reddish gas blasting above the Sun's limb, and even watch them as they slowly change over minutes or hours. It makes an incredible way to appreciate our home star, the source of all the energy that makes life possible on Earth.

Solar Cycle

Solar cycle is a period of about 11 years in which fluctuations in the number and size of sunspots and solar prominences are repeated. Sunspot groups have a magnetic field with a north and a south pole, and, in each 11-year rise and fall, the same polarity leads in a given hemisphere, while the opposite polarity leads in the other. In each rise and fall, the latitude of sunspot eruption starts around 30° and drifts to the equator, but the magnetic fields of the follower spots (sunspots usually come in pairs, called leader and follower) drift poleward and reverse the polar field. In the next 11-year period, the magnetic polarities are reversed but follow the same pattern. Therefore, the magnetic period is 22 years.

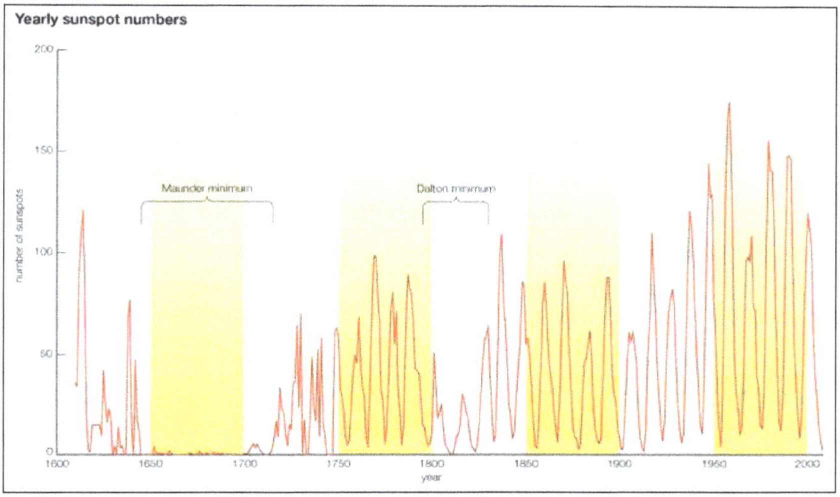

Graph of average yearly sunspot numbers showing the 11-year solar cycle.

Although sunspots were known as early as 1600, no one noticed that their number changed with time until the German amateur astronomer Samuel Heinrich Schwabe announced the 11-year cycle in 1843. The 22-year magnetic cycle was discovered in 1925 by the American astronomer George Ellery Hale.

In 1894 the English astronomer E. Walter Maunder pointed out that very few sunspots were observed between 1645 and 1715, a period now known as the Maunder minimum. This period coincided with the coldest part of the Little Ice Age in the Northern Hemisphere, when the River Thames in England froze over during winter, Viking settlers abandoned Greenland, and Norwegian farmers demanded that the Danish king recompense them for lands occupied by advancing glaciers. The event was confirmed by the American astronomer J.A. Eddy, using carbon isotope ratios in tree rings. During this time the 11-year cycle continued but with a much-reduced amplitude. The data suggest that other such events occurred even earlier in the previous millennium. The late 18th and early 19th centuries also had a brief period of decreased sunspot activity, the Dalton minimum that also coincided with a period that was slightly cooler than normal. The physical mechanism that explains how changes in solar activity affect Earth's climate is unknown, and these episodes, however suggestive, do not prove that lower sunspot numbers produce cooling.

The solar cycle that began in 2008 will reach maximum in 2013, but that maximum is predicted to have only one-half of the number of sunspots seen in the previous cycle. This decrease in the number of sunspots has led some solar physicists to predict an upcoming period of inactivity like the Dalton minimum.

Solar-terrestrial Effects

Besides providing light and heat, the Sun affects Earth through its ultraviolet radiation, the steady stream of the solar wind, and the particle storms of great flares. The near-ultraviolet radiation from the Sun produces the ozone layer, which in turn shields the planet from such radiation. The other effects, which give rise to effects on Earth called space weather, vary greatly. The soft (long-wavelength) X-rays from the solar corona produce those layers of the ionosphere that make short-wave radio communication possible. When solar activity increases, the soft X-ray emission from the corona (slowly varying) and flares (impulsive) increases, producing a better reflecting layer but eventually increasing ionospheric density until radio waves are absorbed and shortwave communications are hampered. The harder (shorter wavelength) X-ray pulses from flares ionize the lowest ionospheric layer (D-layer), producing radio fade-outs. Earth's rotating magnetic field is strong enough to block the solar wind, forming the magnetosphere, around which the solar particles and fields flow. On the side opposite to the Sun, the field lines stretch out in a structure called the magnetotail. When shocks arrive in the solar wind, a short, sharp increase in the field of Earth is produced. When the interplanetary field switches to a direction opposite Earth's field, or when big clouds of particles enter it, the magnetic fields in the magnetotail reconnect and energy is released, producing the aurora borealis (northern lights). Each time a big coronal hole faces Earth, the solar wind is fast, and a geomagnetic storm occurs. This produces a 27-day pattern of storms that is especially prominent at sunspot minimum. Big flares and other eruptions produce coronal mass ejections, clouds of energetic particles that form a ring current around the magnetosphere, which produces sharp fluctuations in Earth's field, called geomagnetic storms. These phenomena disturb radio communication and create voltage surges in long-distance transmission lines and other long conductors.

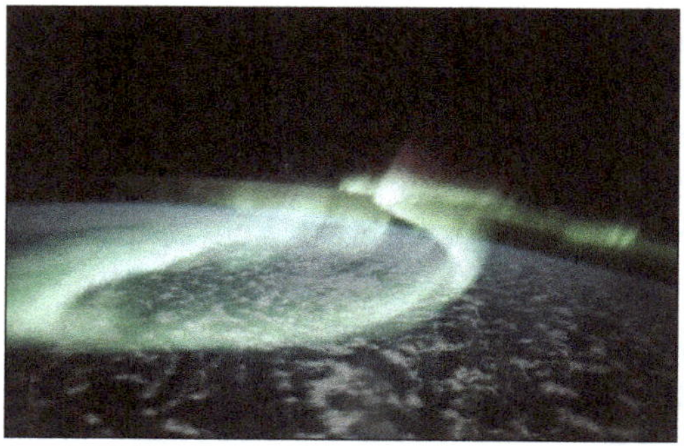

A display of aurora australis, or southern lights, manifesting itself as a glowing loop, in an image of part of Earth's Southern Hemisphere taken from space by astronauts aboard the U.S. space shuttle orbiter Discovery on May 6, 1991. The mostly greenish blue emission is from ionized oxygen atoms at an altitude of 100–250 km (60–150 miles). The red-tinged spikes at the top of the loop are produced by ionized oxygen atoms at higher altitudes, up to 500 km (300 miles).

Perhaps the most intriguing of all terrestrial effects are the possible effects of the Sun on the climate of Earth. The Maunder minimum seems well established, but there are few other clear effects. Yet most scientists believe an important tie exists, masked by a number of other variations.

Because charged particles follow magnetic fields, corpuscular radiation is not observed from all big flares but only from those favourably situated in the Sun's western hemisphere. The solar rotation makes the lines of force from the western side of the Sun (as seen from Earth) lead back to Earth, guiding the flare particles there. These particles are mostly protons because hydrogen is the dominant constituent of the Sun. Many of the particles are trapped in a great shock front that blows out from the Sun at 1,000 kilometres per second. The flux of low-energy particles in big flares is so intense that it endangers the lives of astronauts outside the terrestrial magnetic field.

Solar Magnetic Fields

The Sun has a very large and very complex magnetic field. The magnetic field at an average place on the Sun is around 1 Gauss, about twice as strong as the average field on the surface of Earth (around 0.5 Gauss). Since the Sun's surface is more than 12,000 times larger than Earth's, the overall influence of the Sun's magnetic field is vast.

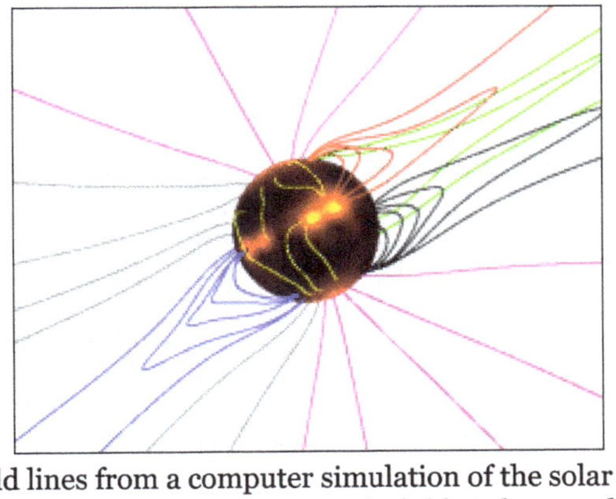

Magnetic field lines from a computer simulation of the solar corona show some of the complexity of the Sun's magnetic field. Colors on the Sun's surface show the strength of the magnetic field (yellow is largest).

The magnetic field of the Sun actually extends far out into space, beyond the furthest planet (Pluto). This distant extension of the Sun's magnetic field is called the Interplanetary Magnetic Field (IMF). The solar wind, the stream of charged particles that flows outward from the Sun, carries the IMF to the planets and beyond. The solar wind and IMF interact with planetary magnetic fields in complex ways, generating phenomena such as the aurora.

Overall, the basic shape of the Sun's magnetic field is like the shape of Earth's field or like the field of a simple bar magnet. However, superimposed on this basic field (called a dipole field) is a much more complex series of local fields that vary over time. Places where the Sun's magnetic field is especially strong are called active regions, and often produce telltale sunspots. The local magnetic field in the neighborhood of a large sunspot can be as strong as 4,000 Gauss much, much greater than the Sun's average field. Disruptions in magnetic fields near active regions can spawn energetic explosions on the Sun such as solar flares and Coronal Mass Ejections. The degree of complexity of the Sun's field waxes and wanes over the course of each sunspot cycle.

The exact nature and source of the Sun's magnetic field are areas of ongoing research. Turbulent motions of charged plasmas in the Sun's convective zone clearly play a role. Some of the Sun's magnetism may even be a remnant from the primordial cloud from which the Sun formed.

Some of the spectacular structures seen in the solar atmosphere, such as solar prominences and coronal loops, are fantastic visible indicators of material flowing along magnetic field lines which arc thousands of kilometers above the Sun's surface.

Solar Wind

The Sun's outer atmosphere, the super-hot corona, is the source of the solar wind, a steady outflow of charged particles from the Sun. These particles have gained enough energy to fill the heliosphere, a region of space that extends well past the orbit of Pluto. As these particles flow past Earth's orbit, they're traveling at an average of 400 kilometers per second. Though the Sun can lose more than a million tons of material each second, the amount is still negligible compared to the Sun's total mass.

The solar wind pushing on Earth's magnetic field.

The solar wind is primarily composed of roughly equal numbers of protons and electrons, as well as a few heavier ions. The particles' velocities are highest over coronal holes, areas near the Sun's poles associated with "open" magnetic field lines that allow material to flow more easily into space. Particles flow out more slowly near the Sun's equator, where magnetic field lines loop back on themselves and trap coronal material.

Earth's magnetic field protects our planet, carving out a cavity in the solar wind called the magnetosphere. Because of the solar wind's pressure on the magnetic field, the magnetosphere is compressed on the Sun-facing side. On the opposite side, it stretches out into a magnetotail.

Occasionally, the Sun's charged particles find their way into Earth's magnetosphere, and spiral along magnetic field lines toward the poles, where they interact with particles in the Earth's upper atmosphere to create auroras.

Solar Flares

A solar flare is basically a giant explosion on the surface of our Sun which occurs when magnetic field lines from sunspots tangle and erupt. A solar flare is defined as a sudden, rapid, and intense variation in brightness. A solar flare occurs when magnetic energy that has built up in the solar atmosphere is suddenly released. Material is heated to many millions of degrees in just minutes and radiation is emitted across virtually the entire electromagnetic spectrum, from radio waves at the long wavelength end, through optical emission to X-rays and gamma rays at the short wavelength end. The amount of energy released is equivalent to millions of nuclear bombs exploding all at the same time! Solar flares are an often occurrence when the Sun is active in the years around solar maximum. Many solar flares can occur on just one day during this period! Around solar minimum, solar flares might occur less than once per week. Large flares are less frequent than smaller ones. Some (mostly stronger) solar flares can launch huge clouds of solar plasma into space which we call a coronal mass ejection. When a coronal mass ejection arrives at Earth, it can cause a geomagnetic storm and intense auroral displays.

A spectacular solar flare in the 193 Ångström wavelength.

Classification of Solar Flares

Solar flares are classified as A, B, C, M or X according to the peak flux (in watts per square metre, W/m2) of 1 to 8 Ångströms X-rays near Earth, as measured by XRS instrument on-board the GOES-15 satellite which is in a geostationary orbit over the Pacific Ocean. The table below shows us the different solar flare classes:

Class	W/m^2 between 1 & 8 Ångströms
A	$<10^{-7}$
B	$\geq 10^{-7}$ $<10^{-6}$
C	$\geq 10^{-6}$ $<10^{-5}$
M	$\geq 10^{-5}$ $<10^{-4}$
X	$\geq 10^{-4}$

Each X-ray class category is divided into a logarithmic scale from 1 to 9. For example: B1 to B9, C1 to C9, etc. An X2 flare is twice as powerful as an X1 flare, and is four times more powerful than an M5 flare. The X-class class category is slightly different and doesn't stop at X9 but continues on. Solar flares of X10 or stronger are sometimes also called "Super X-class solar flares."

A and B-class Solar Flares

The A and B-class are the lowest class of solar flares. They are very common and not very interesting. The background flux (amount of radiation emitted when there are no flares) is often in the B-range during solar maximum and in the A-range during solar minimum.

C-class Solar Flares

C-class solar flares are minor solar flares that have little to no effect on Earth. Only C-class solar flares which are long in duration might produce a produce a coronal mass ejection but they are usually slow, weak and rarely cause a significant geomagnetic disturbance here on Earth. The background flux (amount of radiation emitted when there are no flares) can be in the lower C-class range when a complex sunspot region inhabits the Earth-facing solar disk.

M-class Solar Flares

M-class solar flares are what we call the medium large solar flares. They cause small (R1) to moderate (R2) radio blackouts on the daylight side of the Earth. Some eruptive M-class solar flares can also cause solar radiation storms. Strong, long duration M-class solar flares are likely candidates to launch a coronal mass ejection. If the solar flare takes place near the center of the Earth-facing solar disk and launches a coronal mass ejection towards our planet, there is a high probability that the resulting geomagnetic storm is going to be strong enough for aurora on the middle latitudes.

X-class Solar Flares

X-class solar flares are the biggest and strongest of them all. On average, solar flares of this magnitude occur about 10 times a year and are more common during solar maximum than solar minimum. Strong to extreme (R3 to R5) radio blackouts occur on the daylight side of the Earth during the solar flare. If the solar flare is eruptive and takes place near the center of the Earth-facing solar disk, it could cause a strong and long lasting solar radiation storm and release a significant coronal mass ejection that can cause severe (G4) to extreme (G5) geomagnetic storming at Earth.

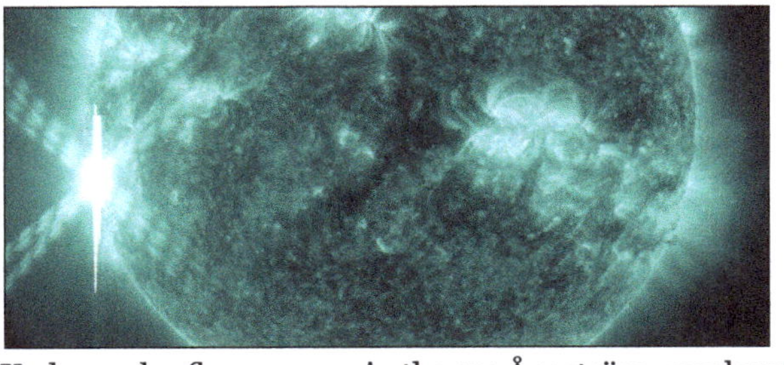

An X-class solar flare as seen in the 131 Ångström wavelength.

So what's above X9? The X-class continues after X9 instead of getting a new letter and these solar flares are often referred to as "Super X-class" solar flares. Solar flares that reach or even surpass the X10 class are however very rare and occur only a few times during a solar cycle. It is actually a good thing that these powerful solar flares do not occur so often as the consequences on Earth could be severe. The coronal mass ejections which can be launched by such solar flares are known to be able to cause issues with our modern technology like satellites and power lines.

One thing to note with super X-class flares is that an X20 solar flare is not 10 times as strong as an X10 solar flare. An X10 solar flares equals an X-ray flux of 0.001 Watts/m^2 while an X20 solar flare equals 0.002 Watts/m^2 in the 1-8 Ångstrom wavelength.

The largest solar flare ever recorded since satellites started to measure them in 1976 was estimated to be an X28 solar flare which occurred on November 4th, 2003 during Solar Cycle 23. The XRS long channel on the GOES-12 satellite was saturated at X17 for 12 minutes by the intense radiation. A later analysis of the available data yield an estimated peak flux of X28 however there are scientists who think that this solar flare was even stronger than X28. A good thing for us was that the sunspot group which produced this solar flare had already rotated largely of the Earth-facing solar disk when the X28 solar flare occurred. A thing to note is that there has not been a solar flare that saturated the XRS channels on GOES-15 as of March 2017 but it is expected that it will saturate at about the same flux levels.

High Frequency (HF) Radio Blackouts caused by Solar Flares

Bursts of X-ray and Extreme Ultra Violet radiation which are emitted during solar flares and can cause problems with High Frequency (HF) radio transmissions on the sunlit side of the Earth and are most intense at locations where the Sun is directly overhead. It is mostly High Frequency (HF) (3-30 MHz) radio communication that is affected during such events, although fading and diminished reception may spill over to Very High Frequency (VHF) (30-300 MHz) and higher frequencies.

These blackouts are a result of enhanced electron densities in the lower ionosphere (D-layer) during a solar flare which causes a large increase in the amount of energy radio waves lose when it passes through this layer. This process prevents the radio waves from reaching the much higher E, F1 and F2 layers where these radio signals normally refract and bounce back to Earth.

Radio blackouts caused by solar flares are the most common space weather events to affect Earth and also the fastest to affect us. Minor events occur about 2000 times each solar cycle. The electromagnetic emission produced during flares travels at the speed of light taking just over 8 minutes to travel from the Sun to Earth. These types of radio blackouts can last from several minutes to several hours depending on the duration of the solar flare. How severe a radio blackout is depends on the strength of the solar flare.

The Highest Affected Frequency (HAF) during an X-ray radio blackout during local noon is based on the current X-ray flux value between the 1-8 Ångström. The Highest Affected Frequency (HAF) can be derived by a formula. Here is a table where you can see what the Highest Affected Frequency (HAF) is during a specific X-ray flux.

GOES X-ray class & flux	Highest Affected Frequency
M1.0 (10^{-5})	15 MHz
M5.0 (5×10^{-5})	20 MHz
X1.0 (10^{-4})	25 MHz
X5.0 (5×10^{-4})	30 MHz

R-scale

NOAA uses a five-level system called the R-scale, to indicate the severity of an X-ray related radio blackout. This scale ranges from R1 for a minor radio blackout event to R5 for an extreme radio blackout event; with R1 being the lowest level and R5 is the highest level. Every R-level has a certain X-ray brightness associated with it. This ranges from R1 for an X-ray flux of M1 to R5 for an X-ray flux of X20. On Twitter we provide alerts as soon as a certain radio blackout threshold has been reached. Because each blackout level represents certain GOES X-ray brightness, you can associate these alerts directly with a solar flare that is occurring at that moment. We can define the following radio blackout classes:

R-scale	Description	GOES X-ray threshold by class & flux	Average frequency

R1	Minor	M1 (10^{-5})	2000 per cycle (950 days per cycle)
R2	Moderate	M5 (5×10^{-5})	350 per cycle (300 days per cycle)
R3	Strong	X1 (10^{-4})	175 per cycle (140 days per cycle)
R4	Severe	X10 (10^{-3})	8 per cycle (8 days per cycle)
R5	Extreme	X20 (2×10^{-3})	Less than 1 per cycle

The image below shows the effects of an X1 (R3-strong) solar flare on the sunlit side of the Earth. We can see that the Highest Affected Frequency (HAF) is about 25 MHz there where the Sun is directly overhead. Radio frequencies lower than the HAF suffer an even greater loss.

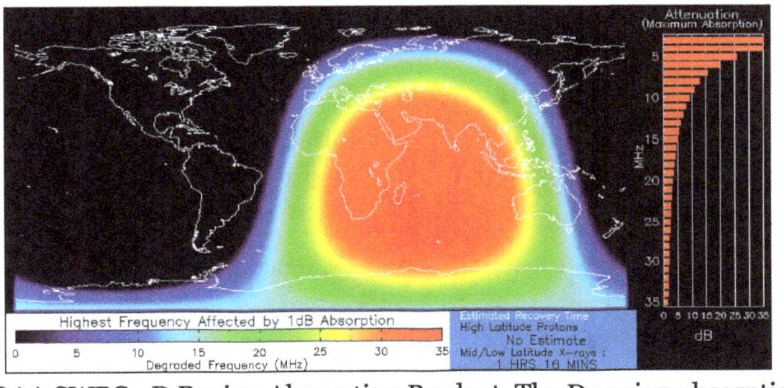

NOAA SWPC - D Region Absorption Product. The D-region absorption prediction model is used as a guide to understand the high frequency (HF) radio degradation and communication interruptions that this can cause.

References

- Solar-atmosphere, place, Sun-54142: britannica.com, Retrieved 29 August, 2019
- What-are-sunspots: spaceweatherlive.com, Retrieved 14 May, 2019
- What-is-a-solar-prominence, sky-events, observing: astronomy.com, Retrieved 05 June, 2019
- Solar-terrestrial-effects, place, Sun: britannica.com, Retrieved 10 March, 2019
- Solar-wind, astronomy-resources: skyandtelescope.com, Retrieved 27 July, 2019
- What-are-solar-flares: spaceweatherlive.com, Retrieved 08 March, 2019

4

Black Hole: A Comprehensive Study

A black hole is a region in the space which exhibits very strong gravitational acceleration. There are diverse concepts related to black hole such as physics of black holes, black hole entropy, black hole thermodynamics, astrophysics of black holes, creation of a black hole, etc. This chapter has been carefully written to provide an easy understanding of these aspects of black hole.

Black holes are regions of space in which gravitational fields are so strong that no particle or signal can escape the pull of gravity. The boundary of this no-escape region is called the event horizon, since distant observers outside the black hole cannot see (cannot get light from) events inside.

Artist's conception of an accretion disk around a black hole.

Although the fundamental possibility of such an object exists within Newton's classical theory of gravitation, Einstein's theory of gravity makes black holes inevitable under some circumstances. Prior to the early 1960s, black holes seemed to be only an interesting theoretical concept with no astrophysical plausibility, but with the discovery of quasars in 1963 it became clear that very exotic astrophysical objects could exist. Nowadays it is taken for granted that black holes do exist in at least two different forms. Stellar mass black holes are the endpoint of the death of some stars, and supermassive black holes are the result of coalescences in the centers of most galaxies, including our own.

No signal can propagate from inside a black hole, but the gravitational influence of a black hole is always present. (This influence does not *propagate* out of the hole; it is permanently present outside, and depends only on the total amount of mass, angular momentum, and electric charge that have gone into forming the hole.) Black holes can be detected through the influence of this strong gravity on the surroundings just outside the hole. In this way, stellar mass holes produce detectable X-rays, supermassive black holes produce a wide spectrum of electromagnetic signals, and both types can be inferred from the orbital motion of luminous stars and matter around them. Phenomena involving black holes of any mass can produce strong gravitational waves, and are of interest as sources for present and future gravitational wave detectors.

Classical vs. Relativistic Black Holes

Something like a black hole exists within Newton's classical theory of gravity. In that theory, an energy argument tells us that there is an escape velocity $v_{esc} = \sqrt{2GM/R}$ from the surface of any spherical object of mass M and radius R. If this velocity is greater than the speed of light c then light from this object cannot escape to infinity. Thus the condition for such an "unseeable" object is,

$$R<2GM/c^{2}.$$

In the classical theory, a particle could overcome this gravity with strong enough engines to provide the energy needed for escape. This is not so in general relativity, Einstein's theory of gravitation. In that theory, escaping the black hole is equivalent to moving faster than light, an impossibility in relativity.

To understand the relativistic black hole it is useful to think of space being dragged inward towards a gravitational center, at a faster rate near the center than far from it. The distance at which space is moving inward at the speed of light represents the location of the event horizon, since no signal can progress outward through space faster than c. This comparison is more than a metaphor; black hole analog experiments with accelerating gas flows and other phenomena are being designed.

An important difference from Newton's theory is that Einstein's, and other relativistic theories of gravitation are nonlinear in the sense that gravitation (as well as mass) can be a source of gravity. Thus when a massive object collapses small enough, the tendency to continue the collapse and form a black hole can become unstoppable.

Stationary Black Holes

In the Newtonian theory, gravity is described by the potential Φ. Inside a spherical object the form of $\Phi(r)$ depends on the interior structure, but in the vacuum outside matter the potential $-GM/r$ depends only on the interior mass. Similarly, in Einstein's theory the stationary (time-independent) spherically symmetric exterior solution, called the Schwarzschild spacetime, depends only on the mass of the interior object. If the interior object is small enough, then the Schwarzschild exterior extends to small enough radius that there is a horizon, a surface across which light cannot move outward. This horizon radius $R_H=2GM/c^2$ is, coincidentally, the same as the critical radius for "unseeable" objects in Newton's theory. (The meaning of "radius" as distance to the center is not straightforward for the Schwarzschild solution. Radius RH here actually means that the area of the event horizon is $4\pi R^2H$).

In Einstein's theory, the "exterior" solution can be taken to apply with no interior solution. In this case it is gravity itself, rather than matter, that acts as the source of gravity. The inward-extended exterior solution does not reach a center, but rather is connected via a spacetime bridge to another universe, or another section of our own. For an astrophysical black hole, formed from the collapse of matter, a physical solution for the matter distribution replaces the pure vacuum Schwarzschild solution in the interior of the black hole. This physical solution lacks the spacetime bridge of ideal mathematical black holes, but contains a central "singularity" where matter is compressed to infinite density. Very close to this singularity it is expected that the laws of general relativity will no longer apply, and as-yet unknown laws of quantum gravity are needed.

A more general stationary black hole solution of Einstein's theory is the Kerr solution, a vacuum spacetime with both mass and angular momentum, and taken to represent a rotating black hole. In its pure mathematical form the Kerr hole contains a spacetime bridge, but as in the case of the Schwarzschild black hole this bridge is absent in realistic black holes that form by the collapse of matter.

Unlike the Schwarzschild spacetime, the Kerr solution is *not* the exterior spacetime of a material object with angular momentum. (In fact no realistic solution has been found to join a Kerr exterior to a material interior.) The Kerr solution only becomes the exterior spacetime asymptotically at very late times after the collapse of an object.

Two other exact mathematical black hole solutions are the Reissner-Nordström spacetime, representing a hole with mass and electrical charge, and the Kerr-Newman spacetime, representing a hole with mass, electrical charge, and angular momentum. These spacetimes are not astrophysically relevant, since astrophysical bodies have negligible net electrical charge.

All of these spacetimes, including the ones with angular momentum, are stationary: that is, they are independent of time. But in relativity there is no unique meaning to time, so an important question is: "Just what 'time' is it of which the stationary black holes are independent?" The answer lies in the fact that one can assign every spacetime point four coordinates, four labels that uniquely identify the location of each point. One of these coordinates is called the "coordinate time." Spacetimes that are said to be stationary, like the spacetime of a Kerr hole, have a special property: the time coordinate may be chosen so that the spacetime geometry is the same at any moment of this time coordinate.

At large distances from a stationary black hole, where spacetime curvatures are weak, this stationary time coordinate can be chosen also to have another important property: to agree with the "proper time," or ordinary clock time, of an observer at rest with respect to the hole. Since we ourselves are more-or-less at rest (or are at nonrelativistic velocities) very far from black holes, this kind of stationary coordinate time is the time used in astronomical observations. For observers near the hole, however, proper time and stationary coordinate time will not agree. An interval of proper time between two events is shorter (near the horizon, much shorter) than the interval of stationary coordinate time between those two events. The relationship between stationary coordinate and proper time is further complicated by special relativistic time dilation for observers who are rapidly moving. Figure illustrates this by showing the coordinate time vs. radius for a particle falling into a black hole, and comparing it with the proper time measured by an observer riding along with the infalling particle. The progress as measured with proper time is in no way special at the horizon: falling observers will not notice anything strange as they pass the point of no return. But as described in coordinate time, the observer (and likewise, the surface of a collapsing star) takes an infinite amount of time to reach the horizon. Since coordinate time is the proper time of distant observers, astronomers will see the particle reach the horizon only in the infinite future.

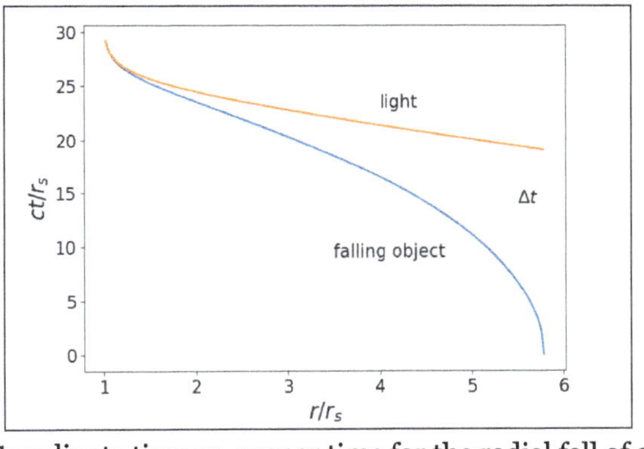

Coordinate time vs. proper time for the radial fall of a particle towards a Schwarzschild black hole.

The two (or more) types of time are sometimes a source of confusion in discussions of black hole phenomena, since they often give totally different answers to the question "how long does it take?"

Black Hole Parameters

Astrophysical black holes are characterized by two parameters: their mass and their angular momentum (or *spin*). The mass parameter M is equivalent to a characteristic length GM/c^2=1.48km(M/M_0), or a characteristic timescale GM/c^3=4.93×10−6sec(M/M_0), where Mo denotes the mass of the Sun. These scales, for example, give the order of magnitude of the radii and periods of near-hole orbits. The timescale also applies to the process in which a developing horizon settles into its asymptotically stationary form. For a stellar mass hole this is of order 10^{-5}sec, while for a supermassive hole of $10^8 M_0$ it is thousands of seconds.

For Schwarzschild holes, and approximately for Kerr holes, the horizon is at radius R_H=$2GM/c^2$. At the horizon, the "acceleration of gravity" has no meaning, since a falling observer cannot stop at the horizon to be weighed. What are relevant at the horizon is the tidal stresses that stretch and distort the falling observer. This tidal stretching is given by the same expression, the gradient of the gravitational acceleration, as in Newtonian theory:

$$2GM/R^3H=c^6/(4G^2M^2)$$

In the case of a solar mass black hole the tidal stress (acceleration per unit length) is enormous at the horizon, on the order of $3\times10^9(M_0/M)^2s^{-2}$: that is, a person would experience a differential gravitational field of about 109 Earth gravities, enough to rip apart ordinary materials. For a supermassive hole, by contrast, the tidal force at the horizon is smaller by a typical factor 10^{10-16} and would be easily survivable. However, at the central singularity, deep inside the event horizon, the tidal stress is *infinite*.

In addition to its mass M, the Kerr spacetime is described with a spin parameter a defined by the dimensionless expression:

$$\frac{a}{M}=\frac{cJ}{GM^2}$$

Where, J is the angular momentum of the hole. For the Sun (based on surface rotation) this number is about 0.2, and is much larger for many stars. Since angular momentum is ubiquitous in astrophysics, and since it is expected to be approximately conserved during collapse and black hole formation, astrophysical holes are expected to have significant values of a/M from several tenths up to and approaching unity.

The value of a/M can be unity (an "extreme" Kerr hole), but it cannot be greater than unity. In the mathematics of general relativity, exceeding this limit replaces the event horizon with an inner boundary on the spacetime where tidal forces become infinite. Because this singularity is "visible" to observers, rather than hidden behind a horizon, as in a black hole, it is called a naked singularity. Toy models and heuristic arguments suggest that as a/M approaches unity it becomes more and more difficult to add angular momentum. The conjecture that such mechanisms will always keep a/M below unity is called cosmic censorship.

The inclusion of angular momentum changes details of the description of the horizon, so that, for example, the horizon area becomes:

$$\text{Horizon area}=(4\pi G^2/c^4)\left[\left(M+\sqrt{M^2}-a^2\right)^2+a^2\right]$$

This modification of the Schwarzschild (a=0) result is not significant until a/M becomes very close to unity. For this reason, good estimates can be made in many astrophysical scenarios with a ignored.

Dynamical Black Holes

The event horizon is defined as the outer boundary of the region from which there is no escape. For stationary black holes this surface is at a fixed location in space, but more generally the horizon is dynamical; it can grow, change shape, oscillate. In particular, a horizon can be born.

The birth of a horizon is a change first studied by Oppenheimer and Snyder for the collapse of spherical pressureless fluid. The general scenario is for a horizon to be born deep inside the collapsing matter and to spread outward.

Small changes in a horizon can be treated as perturbations, greatly simplifying the mathematics of Einstein's equations. In this way, it has been found that black holes have characteristic patterns of oscillations, called quasinormal modes. These modes are like mechanical resonances, but are highly damped by the emission of gravitational waves, and have both periods and damping times on the order of the characteristic time GM/c^3.

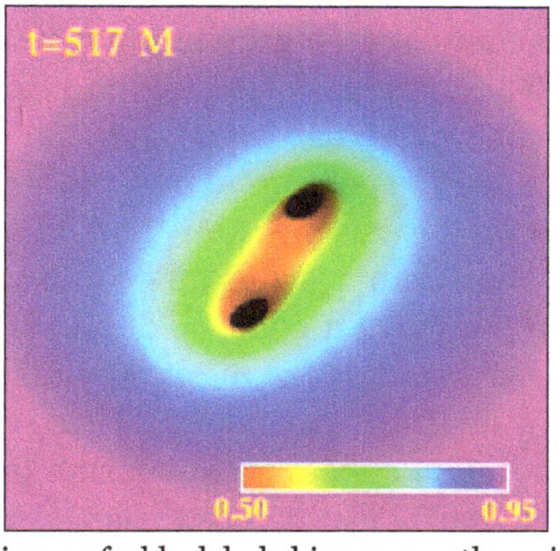

Distorted horizons of a black hole binary near the point of merger.

For large nonspherical changes in a dynamical horizon, only supercomputer simulations can give quantitative answers. The focus of such work has been the merger of two black holes in binary orbit around each other, a scenario of special interest as a source of strong gravitational wave emission. Only in 2005 were technical problems first overcome so that an accurate picture could emerge of how two horizons join to become a single final horizon.

There is one change that a horizon cannot make according to classical (nonquantum) general relativity: it cannot decrease its area. But considerations by Stephen Hawking and others of quantum effects in black hole spacetimes suggest that radiation arising in the close exterior of the black hole can carry off energy, and decreases the mass (and hence horizon area) of the hole. Although no quantum theory of relativistic gravitation currently exists, it is generally accepted that this Hawking radiation will be a feature of any such theory. The radiation behaves as if the horizon were a blackbody (perfect thermal emitter) at a temperature $6\times10^{-8}(Mo/M)$K. Thus for astrophysical black holes, mass loss by Hawking radiation is much less important than the mass increase due to absorption of the 3K cosmic microwave background, and that mass increase is itself negligible.

Astrophysical Black Holes

Black holes in our Universe can be grouped as: primordial black holes, stellar-mass black holes, and supermassive black holes. The first is highly speculative; the second and third are broadly accepted.

Primordial black holes of all masses are postulated to have formed from quantum fluctuations in the early Universe, but those with mass less than around 1013kg would have already evaporated due to Hawking radiation. No significant observational evidence has yet been found of the existence of these objects.

Stellar-mass black holes, ranging from a few to a few tens of solar masses, are normal but rare endpoints in the evolution of massive stars. When a star exhausts its nuclear fuel and cools, it must collapse unless it is supported by nonthermal forces. It is known that such nonthermal forces cannot resist gravitational compression for masses greater than around 1.5 to $3M_{0.}$ (The uncertainty is due to uncertainty in our knowledge of nuclear physics at high densities.) There are many stars far more massive than this, but in their death throes massive stars expel a great deal of their mass in supernova explosions. Even stars initially as massive as 20-30M_0 may blow off enough mass to leave a remnant neutron star smaller than 1.5M_0. Stars more massive than 20-30M_0 may form a neutron star core, only to have it collapse due to fallback of material from the stellar mantle. Still more massive stars (above ~40M_0) may form a black hole directly in collapse, with or without a supernova explosion. These masses are rather uncertain at present, as our understanding of supernova explosion mechanisms is still evolving. Collapsing gas clouds larger than ~100M_0 are expected to dissipate from radiation pressure, which would prevent more massive stars from forming, and consequently sets an approximate upper limit for stellar mass black holes.

Observational evidence for stellar-mass black holes comes primarily from X-ray astronomy. A black hole in a close binary orbit can form an accretion disk of matter pulled off a normal stellar companion. In its inspiral, disk material is heated by shearing, and becomes a strong X-ray emitter. If observations reveal a point-like X-ray emitter with a mass inferred from orbital dynamics to be above that possible for a neutron star, it becomes a black hole candidate. At present there are over a dozen such black hole candidates, including the first object to be identified as a black hole, the X-ray source Cygnus X-1 (estimated mass 10±3M_0).

The other class of observationally-supported black holes is that of supermassive black holes, ranging from hundreds to billions of solar masses. Evidence for the existence of black holes at the upper end of this range is overwhelming. By contrast, the existence of black holes with mass roughly of order 102 to 10^4M_0 referred to as "intermediate-mass black holes," is speculative. The supermassive holes may have begun as primordial or stellar-mass black holes, but have grown through absorption of stars or gas, or through mergers with other holes.

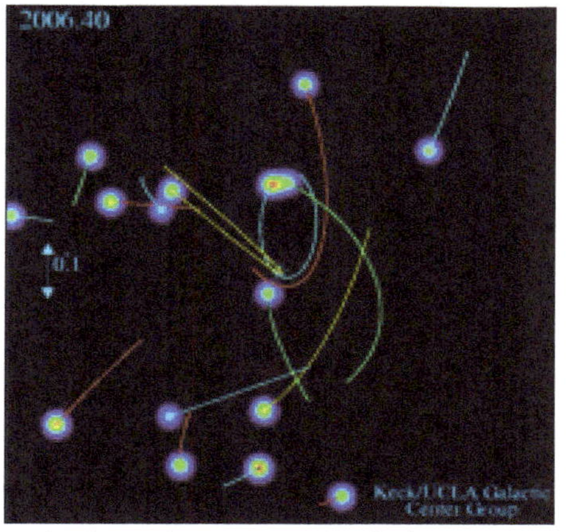

Observed orbits of stars near the dark compact object at the center of our Milky Way galaxy.

Evidence for supermassive black holes originally came from quasars, small but intense radio sources seen at cosmological distances. Their huge luminosities implied high mass, while their rapid variability implied extremely small size. More recently, measurements of Doppler shifts of stars and gas in the centers of galaxies have shown that compact objects of mass greater than $10^6 M_0$ reside in the cores of most galaxies; the very small size of these central objects rules out any plausible alternative to the black hole explanation. The black hole at the center of our own Milky Way galaxy has a measured mass of $\sim 3\times10^6 M_0$, based on the orbits of stars near the radio source Sagittarius A* associated with the Galactic core.

Gravitational Waves from Black Holes

Black holes are of particular interest for gravitational waves, and vice versa. For gravitationally-bound systems, the typical maximum gravitational wave amplitude (dimensionless strain) is $h \sim (2GM/c^2)^2/(RD) \sim R^2H/(RD)$ where D is the distance to the system, M and R are the mass and size of the system, and RH is the horizon radius of a black hole of the same mass. From this it is clear that strongest gravitational waves will involve systems at or near black hole compactness: in particular, systems containing black holes in close proximity to one another.

Conversely, a system containing *only* black holes would not be expected to radiate *any* sort of radiation other than gravitational waves, and in any system containing black holes, gravitational waves provide the most accurate probes of conditions deep in their gravitational wells. In particular, gravitational wave observations could definitively test whether a compact mass is truly a black hole.

The strongest intrinsic source of gravitational waves in the present Universe is the collision of two comparable-massed black holes. A sufficiently tight black hole binary system will lose energy through emission of gravitational waves, causing the orbit to circularize and then slowly shrink over time.

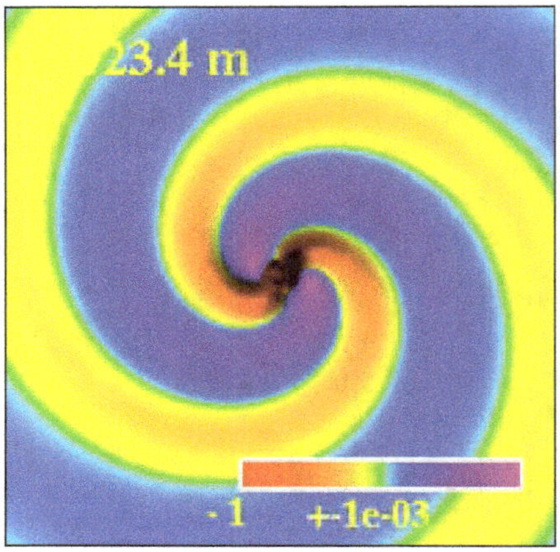

Gravitational waves from a black hole binary inspiral.

When the orbital radius approaches the Schwarzschild radius of the system, complex nonlinear dynamics come into play and the final stages of the inspiral and merger can only be modeled through numerical simulations. These simulations are quite challenging, but recent breakthroughs have led to the first complete inspiral-merger gravitational waveforms. After the black holes have coalesced, they form a single highly-distorted black hole which quickly settles down to a quiescent Kerr state through emission of quasinormal "ringdown" gravitational waves.

Another gravitational emission mechanism for supermassive black holes would come from their capture of compact stellar objects, particularly stellar-mass black holes. Unlike comparable-mass mergers, these captures would be on highly eccentric orbits that would not circularize before merger, resulting in intricate rapidly precessing orbits in the final months or years of inspiral. The consequent gravitational-wave signal would allow for exquisitely precise measurements of the parameters of the supermassive black hole, as well as tests of any deviation from the predictions of general relativity.

For stellar-mass black holes (tens of M_0), the final stages of inspiral and merger occur at frequencies of hundreds to thousands of Hz, a regime targeted by ground-based gravitational-wave observatories. Systems of supermassive black holes that are millions of M_0 emit at frequencies of tens of mHz, which will be detectable by proposed space-based gravitational-wave detectors. Black holes in the billions of M_0 or a stochastic background of signals from more moderate supermassive black holes at earlier stages of their inspiral, might emit waves at nHz frequencies that would be measurable through years-long correlated timing measurements of radio pulsars.

Properties

BHs are the most mysterious objects in the Universe with their uncertainty inside the event horizon. Let's consider the most interesting properties of BHs:

- Stars located near the BH lose some weight in the form of stellar winds;

- The existence of BH was predicted by Karl Schwarzschild who coined the term "the Schwarzschild radius";
- The singularity of BH cancels the usual laws of physics and can produce the new Universe;
- BHs stretch the objects that are close to them;
- BHs not only absorb the stellar wind but also are evaporated; 6. BHs slow the time;
- BHs are the most advanced power plants;
- BHs curve space.

Black Hole Entropy

A black hole may be described as a blemish in spacetime, or a locale of very high curvature. Is it meaningful or desirable to associate entropy with it? Is this possible at all? There are several ways to justify the concept of black hole entropy:

- A black hole is usually formed from the collapse of a quantity of matter or radiation, both of which carry entropy. However, the hole's interior and contents are veiled to an exterior observer. Thus a thermodynamic description of the collapse from that observer's viewpoint cannot be based on the entropy of that matter or radiation because these are unobservable. Associating entropy with the black hole provides a handle on the thermodynamics.
- A stationary black hole is parametrized by just a few numbers: its mass, electric charge and angular momentum (and magnetic monopole charge, except its actual existence in nature has not been demonstrated yet). For any specific choice of these parameters one can imagine many scenarios for the black hole's formation. Thus there are many possible internal states corresponding to that black hole. In thermodynamics one meets a similar situation: many internal microstates of a system are all compatible with the one observed (macro)state. Thermodynamic entropy quantifies the said multiplicity. Thus by analogy one needs to associate entropy with a black hole.
- By blocking all signal travel through it, the event horizon prevents an external observer from receiving information about the black hole. Thus a black hole can be said to hide information. In ordinary physics entropy is a measure of missing information. Hence it makes sense to attribute entropy to a black hole.

Due to the disposition of the local light cone, the event horizon stops any signals bearing interior information from exiting the black hole. Only the hole's mass M angular momentum J and electric charge Q are sensed by an exterior observer.

Formula for Black Hole Entropy

How to express the black hole entropy in a concrete formula? It is clear at the outset that black hole entropy should only depend on the observable properties of the black hole: mass, electric charge and angular momentum. It turns out that these three parameters enter only in the same combination as that which represents the surface area of the black hole. One way to understand why is to recall the "area theorem": *the event horizon area of a black hole cannot decrease; it increases in most transformations of the black hole.* This increasing behavior is reminiscent of thermodynamic entropy of closed systems. Thus it is reasonable that the black hole entropy should be a monotonic function of area, and it turns out to be simplest such function.

If A stands for the surface area of a black hole (area of the event horizon), then the black hole entropy, in dimensionless form, is given by:

$$S_{BH} = \frac{A}{4L^2P} = \frac{c^3 A}{4G\hbar}$$

Where, L_P stands for the Planck length $G\hbar/c^3$ while G, ħ and c denote, respectively, Newton's gravity constant, the Planck-Dirac constant ($h/(2\pi)$) and the speed of light. Of course, if the entropy in the usual (chemist's) form is required, the above should be multiplied by Boltzmann's constant k.

For the spherically symmetric and stationary, or Schwarzschild, black hole, the only parameter is the black hole's mass M, the horizon's radius is $rh=2GM/c^{2,}$ and its area is naturally given by $4\pi r^2 h$, or

$$A = 16\pi(GM / c^2)^2$$

Note that a one-solar mass Schwarzschild black hole has an horizon area of the same order as the municipal area of Atlanta or Chicago. Its entropy is about 4×10^{77}, which is about twenty orders of magnitude larger than the thermodynamic entropy of the sun. This observation underscores the fact that one should not think of black hole entropy as the entropy that fell into the black hole when it was formed.

For the most general type of stationary black hole, the Kerr-Newman black hole (rotating black hole), the hole's parameters are mass *M*, electric charge *Q* and angular momentum *J*, and the horizon is no longer spherical. Nevertheless, in the popular Boyer-Lindquist coordinates $\{t, r, \theta, \varphi\}$ it lies at the fixed radial coordinate:

$$r = r_h \equiv GM / c^2 + \sqrt{(GM / c^2)^2 - (G^{1/2}Q / c^2)^2} - (J / Mc)^2)$$

Consequently the horizon area is given by:

$$A = \int_0^{\pi} d\theta \int_0^{2\pi} d\varphi \sqrt{g\theta\theta g\varphi\varphi} = 4\pi(r_h^2 + (J / Mc)^2)$$

In ordinary statistical mechanics, the entropy *S* is a measure of the multiplicity of microstates that hide behind one particular macrostate. A special case of this is Boltzmann's famous formula $S = \ln W$ where *W* stands for the number of equally probable microstates of a particular macrostate. Since black hole entropy plays a role quite analogous to that of ordinary entropy, e.g. it participates in the second law, many have wondered what the microstates that are counted by black hole entropy are. A partial list of interpretations of black hole entropy is the following:

- Black hole entropy counts the number of internal states of matter and gravity. The perception that a particular black hole (specific *M*, *J* and *Q*) can be formed in many ways originally suggested the notion of black hole entropy; in this approach the internal states of matter and/or gravity are the sought for microstates. Examples of this viewpoint are provided by Frolov and Novikov and by Mukhanov.

- Black hole entropy is the entropy of entanglement between degrees of freedom inside and outside the horizon. For a black hole formed by collapse, the quantum field degrees of freedom external to the horizon should be entangled with those inside it. To an external observer the last are not accessible, so the meaningful state would be the one from which internal degrees of freedom have been traced out (removed). Even if the global state was pure and thus entropy free, this so reduced state will be a mixed one and have entropy associated with it. Calculation that this entanglement entropy is proportional to the horizon area, just as required to explain black hole entropy, but the coefficient is ultraviolet divergent and thus requires renormalization by physical arguments.

- Black hole entropy counts the number of horizon gravitational states. It has been suggested that the sought for states are the states of the gravitational degrees of freedom residing on the black hole's horizon. An example of this approach is a calculation by Carlip based on the group of symmetries at the horizon. It reproduces formula.

- Black hole entropy is a conserved quantity connected with coordinate invariance of the gravitational action. This abstract approach has been championed by Wald as a road to black hole entropy in more general theories of gravity than general relativity. Technically, black hole entropy is the Noether charge of the diffeomorphism symmetry. This notion reproduces formula when the gravitational action is of first order in the curvature, but gives a modified formula for higher order gravity theories.
- Black hole entropy is thermal entropy of the gas of quanta constituting the thermal atmosphere of the black hole. The atmosphere concept comes from Thorne and Zurek and 't Hooft; the last introduced the concept of a "brick wall" to keep the said atmosphere from contacting the horizon, and thus making the entropy infinite. This approach recovers the proportionality of entropy to horizon area, but the coefficient has to be chosen by hand.
- Black hole entropy counts the number of states or excitations of a fundamental string. Strings in string theory have a variety of excitations, so there is a multitude of string states. Therefore, a string has entropy, which turns out to be proportional to its *mass*. This is quite in contrast with black hole entropy. However, an argument by Bowick, Smolin and Wijewardhana suggests that by *adiabatically* (i.e. sufficiently slowly) reducing the string coupling constant g, it is possible to shrink a black hole's size as well as to reduce its mass (while keeping its entropy constant) until eventually it gets to be the size of the string length scale *ls* when the black hole should not be distinguishable from a string. At the corresponding value of g, string and black hole entropy are quite similar. This has been taken to mean that there is a one-to-one correspondence between black hole and string states, where both entities have the same entropy. This picture has been corroborated in the context of five-dimensional extreme black holes. Hence black hole entropy can be understood in terms of string entropy.
- Black hole entropy is equivalent to the thermal entropy of the radiation residing on the boundary of the spacetime containing the black hole. The AdS/CFT correspondence is a mapping between gravitational degrees of freedom of a certain spacetime and the matter (or field) degrees of freedom residing on its boundary. In particular, certain string theories in five dimensional Anti-deSitter (AdS) spacetime are so mapped to conformal field theories on the corresponding spacetime's four-dimensional boundary which bears some resemblance to Minkowski spacetime. Witten has shown that the entropy of a black hole residing in the bulk Anti-deSitter spacetime equals that of thermal radiation of the fields residing on its boundary.

Black Hole Thermodynamics

In the 1800s scientists studying things like heat and the behavior of low density gases developed a theory known as thermodynamics. As the name suggests, this theory describes the dynamic behavior of heat (or more generally energy). The core of thermodynamics is embodied by its four basic laws.

The zeroth law states that if object A is in thermodynamic equilibrium with object B (meaning no net energy flows between them), and object C is in thermodynamic equilibrium with B, then A and C are in thermodynamic equilibrium with each other. Since objects in thermodynamic equilibrium have the same temperature, another way to state this law is that if A has the same temperature as B, and C has the same temperature as B, then A and C have the same temperature. When you put it that way it seems quite obvious, which is why it isn't known as the first law. The other laws were developed first, and as they were refined it became clear the zeroth law should be included as a physical property, not just an assumption.

The first law states that energy is conserved. Since heat is a form of energy, this means an object that is heating up must be getting energy from somewhere. Likewise, if an object is cooling down, the energy it loses must be gained by something else. Conservation of energy was known before thermodynamics, but this law recognized heat as a form of energy.

The second law is perhaps the most misunderstood law of thermodynamics. In its simplest form it can be summarized as "heat flows from hot objects to cold objects". But the law is more useful when it is expressed in terms of entropy. In this way it is stated as "the entropy of a system can never decrease." Many people interpret entropy as the level of disorder in a system, or the unusable part of a system. That would mean things must always become less useful over time, which is why evolution skeptics often claim it violates the second law of thermodynamics.

But entropy is really about the level of information you need to describe a system. An ordered system (say, marbles evenly spaced in a grid) is easy to describe because the objects have simple relations to each other. On the other hand, a disordered system (marbles randomly scattered) take more information to describe, because there isn't a simple pattern to them. So when the second law says that entropy can never decrease, it is say that the physical information of a system cannot decrease. In other words, information cannot be destroyed.

The third law basically states that at absolute zero an object is at its minimum possible entropy (often taken as zero). One consequence of this law is that you cannot cool an object to absolute zero.

classical black holes have "no hair", meaning that they are simply described by their mass, charge and rotation. Because of this, you could toss an object (with a great deal of entropy) into a black hole, and the entropy would simply go away. In other words, the entropy of the system would get smaller, which would violate the second law of thermodynamics. Another way of looking at it would be that the classical black hole has a temperature of absolute zero. This means you could take some hot mass and collapse it into a black hole, which would essentially be cooling an object to absolute zero, in violation of the third law of thermodynamics.

Of course, this ignores the effects of quantum mechanics. When we take quantum mechanics into account, black holes can emit light and other particles through a process known as Hawking radiation. Since a "quantum" black hole emits heat and light, it therefore has a temperature. This means black holes are subject to the laws of thermodynamics.

Integrating general relativity, quantum mechanics and thermodynamics into a comprehensive description of black holes is quite complicated, but the basic properties can be expressed as a fairly simple set of rules known as black hole thermodynamics. Essentially these are the laws of thermodynamics re-expressed in terms of properties of black holes.

The zeroth law states that a simple, non-rotating black hole has uniform gravity at its event horizon. This is kind of like saying that such a black hole is at thermal equlibrium.

The first law relates the mass, rotation and charge of a black hole to its entropy. The entropy of a black hole is then related to the surface area of its event horizon.

The second law again states that the entropy of a black hole system cannot decrease. One consequence of this is that when two black holes merge, the surface area of the merged event horizon must be greater than the surface areas of the original black holes.

The third law states that "extreme" black holes (those with a maximum possible rotation or charge) would have minimum entropy. This means that it would never be possible to form an extreme black hole. For example, it would never be possible to spin a black hole so fast that it would break apart.

The advantage of black hole thermodynamics is that provides a way to get a handle on the complex interactions black holes can have. Thermodynamic black holes have not just mass, charge and rotation, but also temperature and entropy. The rules first devised to describe the heating and cooling of simple gases also seems to apply to black holes.

Structure of a Black Hole

Event Horizon

Event horizon is the boundary marking the limits of a black hole. At the event horizon, the escape velocity is equal to the speed of light. Since general relativity states that nothing can travel faster than the speed of light, nothing inside the event horizon can ever cross the boundary and escape beyond it, including light. Thus, nothing that enters a black hole can get out or can be observed from outside the event horizon. Likewise, any radiation generated inside the horizon can never escape beyond it. For a nonrotating black hole, the Schwarzschild radius delimits a spherical event horizon. Rotating black holes have distorted, nonspherical event horizons. Since the event horizon is not a material surface but rather merely a mathematically defined demarcation boundary, nothing prevents matter or radiation from entering a black hole, only from exiting one. Though black holes themselves may not radiate energy, electromagnetic radiation and matter particles may be radiated from just outside the event horizon via Hawking radiation.

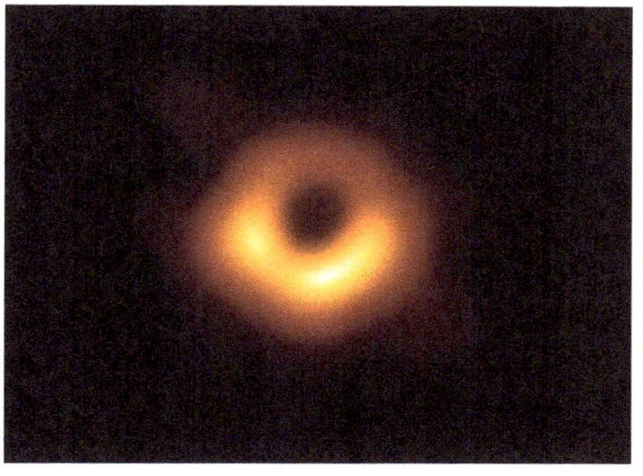

Black hole at the centre of the massive galaxy M87, about 55 million light-years from Earth, as imaged by the Event Horizon Telescope (EHT). The black hole is 6.5 billion times more massive than the Sun. This image was the first direct visual evidence of a supermassive black hole and its shadow. The ring is brighter on one side because the black hole is rotating, and thus material on the side of the black hole turning toward Earth has its emission boosted by the Doppler effect. The shadow of the black hole is about five and a half times larger than the event horizon, the boundary marking the black hole's limits, where the escape velocity is equal to the speed of light.

Singularity

In the center of a black hole is a gravitational singularity, a one-dimensional point which contains a huge mass in an infinitely small space, where density and gravity become infinite and space-time curves infinitely, and where the laws of physics as we know them cease to operate. As the eminent American physicist Kip Thorne describes it, it is "the point where all laws of physics break down".

Current theory suggests that, as an object falls into a black hole and approaches the singularity at the center, it will become stretched out or "spaghettified" due to the increasing differential in gravitational attraction on different parts of it, before presumably losing dimensionality completely and disappearing irrevocably into the singularity. An observer watching from a safe distance outside, though, would have a different view of the event. According to relativity theory, they would see the object moving slower and slower as it approaches the black hole until it comes to a complete halt at the event horizon, never actually falling into the black hole.

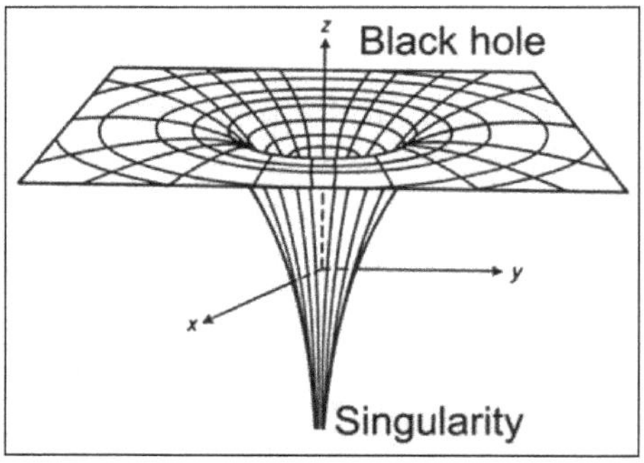

A gravitational singularity is hidden within a black hole.

The existence of a singularity is often taken as proof that the theory of general relativity has broken down, which is perhaps not unexpected as it occurs in conditions where quantum effects should become important. It is conceivable that some future combined theory of quantum gravity (such as current research into superstrings) may be able to describe black holes without the need for singularities, but such a theory is still many years away.

According to the "cosmic censorship" hypothesis, a black hole's singularity remains hidden behind its event horizon, in that it is always surrounded by an area which does not allow light to escape, and therefore cannot be directly observed. The only exception the hypothesis allows (known as a "naked" singularity) is the initial Big Bang itself.

It seems likely, then, that, by its very nature, we will never be able to fully describe or even understand the singularity at the center of a black hole. Although an observer can send signals into a black hole, nothing inside the black hole can ever communicate with anything outside it, so its secrets would seem to be safe forever.

Photon Sphere

A photon sphere is a spherical surface round a non-rotating black hole (or other extremely compact spherically symmetric body) containing all the possible closed orbits of a photon.

All such orbits are circular and unstable. The radius of the photon sphere is 3M, where $M=Gm/c^2$ is the mass-equivalent radius of the body, and m is its mass.

Obviously, there is no photon sphere if the radius of the body is greater than 3M.

For a rotating black hole, there is an outer radius at which the only photon orbit is equatorial circular and retrograde (counter-rotating), and an inner radius at which the only orbit is equatorial circular and prograde (co-rotating).

Bound orbits for a photon exist on each sphere between these two extreme radii. Each such orbit has the approximate shape of a circle which precesses round the sphere between two fixed "latitudes" and with a characteristic angular momentum. This angular momentum increases (becomes more prograde) with decreasing radius, while the fixed "latitude" increases to polar and then decreases again.

Non-rotating Black Hole (Schwarzschild Coordinates)

Radius of event horizon: 2 M

Radius of photon sphere: 3 M

Line element:

$$ds^2 = -\frac{dr^2}{\left(1-\frac{2M}{r}\right)} - r^2 d\theta^2 - r^2 \sin^2\theta d\phi^2 + \left(\left(1-\frac{2M}{r}\right)\right)$$

free-fall equations for mass m, energy parameter E and angular momentum parameter L:

$$\frac{dt}{d\tau} = \frac{E}{\left(1-\frac{2M}{r}\right)}$$

$$\frac{dr}{d\tau} = \frac{L}{r^2}$$

$$\frac{dt}{d\tau} = \pm\sqrt{E^2 - \left(1-\frac{2M}{r}\right)\left(m^2 - \frac{L^2}{r^2}\right)}$$

Rotating Black Hole (with Angular Momentum a M)

Radii of event horizons:

$$R_{\pm} = M \pm \sqrt{M^2 - a^2}$$

radius of innermost photon orbit (prograde equatorial and circular):

$$r_- = 2M\left(1+\cos\frac{2}{3}\cos^{-1}\frac{-a}{M}\right)$$

radius of innermost photon orbit (prograde equatorial and circular):

$$r_+ = 2M\left(1+\cos\frac{2}{3}\cos^{-1}\frac{-a}{M}\right) = 2M\left(1+\cos\left(120^0 - \frac{2}{3}\cos^{-1}\frac{a}{M}\right)\right)$$

The Maths

A body free-falling near a non-rotating black hole follows a trajectory with three constant parameters, m E and L, which may be thought of as its mass energy and angular momentum.

For a photon, m is zero. The usual Schwarzschild coordinates, are related to the "age", τ, of a photon (measured as number of wavelengths, since of course the "proper time" of a photon does not change) by the equations:

$$\frac{dt}{d\tau} = \frac{E}{\left(1 - \frac{2M}{r}\right)}$$

$$\frac{dr}{d\tau} = \frac{L}{r^2}$$

$$\frac{dt}{d\tau} = \pm E\sqrt{1 - \left(1 - \frac{2M}{r}\right)L^2 / E^2 r^2}$$

When, L/E = 3√3M, the last equation is:

$$\frac{dt}{d\tau} = \pm E\sqrt{1 + 6M / r(1 - 3M / r)}$$

from which obviously one solution is the circular motion:

$$r = 3M \ and \ d\theta / dt = 1 / 3\sqrt{3M}$$

Accordingly, a photon with L/E = 3√3M can orbit on the photon sphere (r = 3M) with period 6 π √ 3 M $6\pi\sqrt{3M}$, or can approach the photon sphere, circling ever closer either just outside or just inside it with approximately the same period, but never quite reaching it.

Lens and mirror effects: Similarly, a photon with L/E slightly greater than 3√3M may circle the photon sphere a number of times before returning to distant space. So a black hole can act as a lens giving rise to n ring-shaped images of a background star, each ring corresponding to light which has circled 1, 2, 3, n times around, for some positive integer n (which depends on the distance beyond the black hole).

And it can act as a mirror giving rise to n ring-shaped images of a foreground star, in the same way.

Ergosphere

Ergosphere is a region outside the event horizon, where gravitational forces start to influence objects movements. Objects here can no longer remain stationary in space. Depending on the distance between object and to the event horizon, the influence can be extremely strong or very weak. Close to the event horizon, where escape velocity is nearly the speed of light gravitational will rip objects apart and eventually draw most of their material in. Far away the effects are virtually non-existent. Objects in the Ergosphere can escape the forces of a black hole if their speed is higher than the appropriate escape velocity. General relativity theory predicts that any rotating mass drags surrounding space-time with it. This makes the ergosphere not just a characteristic of black holes, but it is present with all regular cosmic objects of mass, including earth, planets or the sun.

Creation of a Black Hole

A slightly different kind of supernova explosion occurs when even larger, hotter stars (blue giants and blue supergiants) reach the end of their short, dramatic lives. These stars are hot enough to burn not just hydrogen and helium as fuel, but also carbon, oxygen and silicon. Eventually, the fusion in these stars forms the element iron (which is the most stable of all nuclei, and will not easily fuse into heavier elements), which effectively ends the nuclear fusion process within the star. Lacking fuel for fusion, the temperature of the star decreases and the rate of collapse due to gravity increases, until it collapses completely on itself, blowing out material in a massive supernova explosion.

If the mass of the compressed remnant of the star exceeds about 3 - 4 solar masses, then even the degeneracy pressure of neutrons is insufficient to halt the collapse and, instead of forming a neutron star, the core collapses completely into a gravitational singularity, a single point containing all the mass of the entire original star. The gravity in such a phenomenon is so strong that it overwhelms all other forces, to the extent that even light can not escape from it, hence the name black hole. Thus, the gravity of a body just a few times denser than a neutron star would result in its inevitable further collapse into a black hole.

Simulated black hole in front of the Milky Way.

Contrary to popular belief, a black hole does not just "suck up" everything around it in an uncontrolled orgy of destruction: it actually exerts no more gravitational pull on the objects around it than the original star from which it was formed, and any objects orbiting the original star (and which survived the supernova blast) would now orbit a black hole instead (an object would need to approach quite close to a black hole before being sucked in). The very largest blue stars may skip even the supernova stage, so that even their outer shells become incorporated into the singularity.

By definition, we cannot observe black holes directly, but they can be detected by the gravitational effect they exert on other bodies or on light rays. This is especially easy to spot in the case of binary star systems where an ordinary star is orbiting around a black hole. In the early 1990s, Reinhard Genzel pioneered this work, using the then new technique of adaptive optics to plot and track the motions of stars near the center of our own Milky Way galaxy, to show that they must be orbiting a very massive, but invisible, object. From the immense speed with which the stars closest to the center of the galaxy are orbiting - millions of kilometers per hour - we know that there is a "supermassive black hole" (known as Sagittarius A) at the center of the Milky Way, with a mass of around 2 - 4 million times that of our Sun. In addition, in the Milky Way galaxy alone, there are many millions of black holes of at least ten solar masses each.

Supermassive black holes lurk in the centers of most galaxies, forming the hubs around which the galaxies rotate. In fact, from observations of the intense radiation of gases swirling around them at close to the speed of light, we can infer that there are much larger supermassive black holes in the centers of other galaxies, some of them weighing as much as several billion suns. The black hole at the center of a galaxy known as M87 has a mass estimated at around 20 billion solar masses, and may be as large as our entire Solar System.

It seems likely that the early universe, in which very large, short-lived stars were the norm, was scattered with many, many black holes, which gradually merged together over time, creating larger and larger black holes. Observations have shown that is not uncommon for two black holes to swirl around each other in a kind of cosmic dance as their gravitational fields interact. The ripples in space-time caused by two black holes orbiting around each other - typically in a three-leaved clover shape or more complex multi-pass configuration, rather than the simple orbit of an electron within an atom, and ever-smaller and faster as the two objects inevitably approach each other - can be recorded visually and even audibly.

In the case of the largest events, moments after the creation of a black hole, the heat and the hugely amplified magnetic field of the collapsing star combine to focus a pair of tight beams or jets of radiation, perpendicular to the spinning plane of the accretion disk. These beams focus vast amounts of particles and energy (of the order of a billion billion times the energy output of our Sun) away from the black hole at close to the speed of light. The shock waves of this massively energetic beam cause gamma rays to be emitted in a phenomenon known as a "gamma ray burst" or "hypernova" event (so named because its energy and brightness dwarfs even that of a supernova, by a factor of up to a hundred million times). Gamma ray bursts are by far the brightest electromagnetic events occurring in the universe, and can last from mere milliseconds to nearly an hour - a typical burst lasts a few seconds - usually followed by a longer-lived "afterglow" emitting at longer wavelengths (x-ray, ultraviolet, visible, infrared and radio waves). It is likely that collisions between neutron stars, or between a neutron star and a black hole, can also cause gamma ray bursts.

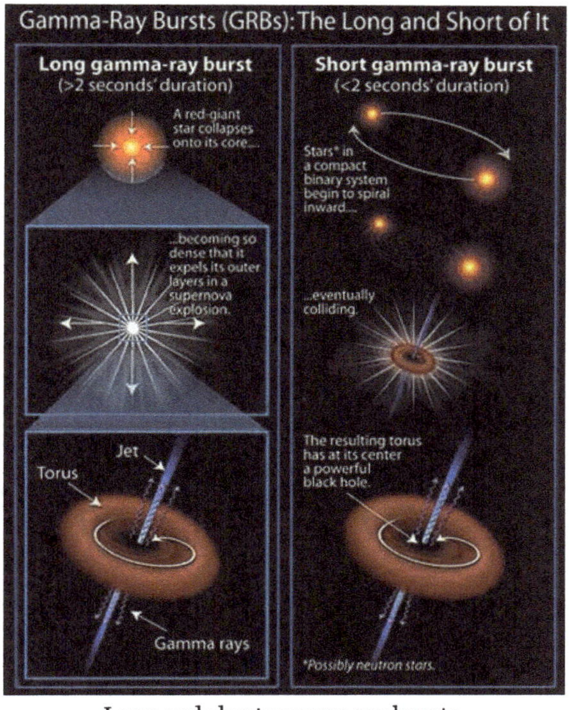

Long and short gamma ray bursts.

Interestingly, it appears to be easier for stars with fewer heavy elements to turn hypernova and generate gamma ray bursts. That, and the fact that larger, more short-lived stars were more common earlier in the life of the universe, means that the phenomenon of gamma ray bursts is actually rarer today than it was. Having said that, NASA's Swift Probe, launched in 2004 with a mission specifically to locate gamma ray bursts throughout the universe, is recording at least one such event each day, so these are not rare incidents. (It should be remembered that any supernovas or gamma ray bursts we observe today in galaxies, say, 9 billion light years away, actually occurred 9 billion years ago.)

References

- Black-hole: scholarpedia.org, Retrieved 12 April, 2019
- Properties-of-black-holes-and-their-searches-at-the-LHC-315792274: researchgate.net, Retrieved 30 August, 2019
- Bekenstein-Hawking-entropy: scholarpedia.org, Retrieved 20 March, 2019
- Event-horizon-black-hole: britannica.com, Retrieved 29 July, 2019
- Threads, what-is-a-photon-sphere-762998: physicsforums.com, Retrieved 16 June, 2019
- Blackholes-blackholes: physicsoftheuniverse.com, Retrieved 13 July, 2019

5

Astronomical Tools and Instruments

Astronomical tools and instruments are used to perform several astronomical observations and calculations. They include optical telescope, solar telescope, radio telescope, hubble space telescope, etc. This chapter closely examines these astronomical tools and instruments to provide an easy understanding of the subject.

Optical Telescope

Optical telescopes have undergone several changes since their invention in the late sixteenth century. All telescopes gather light from a large area and bring it to a common focus. But the way they focus the light varies with design.

The earliest telescopes - like the ones Galileo used are the simplest: we call them refractors. They consist of a long, narrow tube with a lens at the front end. Light which passes through the lens is bent, so that initially parallel rays meet near the bottom of the tube.

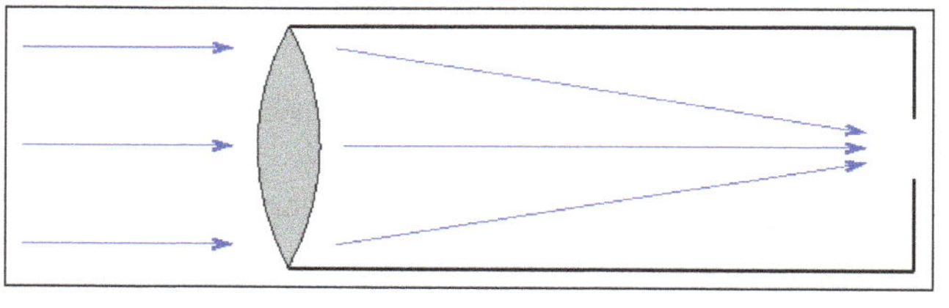

Refractors are easy to make and, when small, relatively inexpensive. Large telescopes of this sort become unwieldy. The largest refractor ever put to practical use is the Yerkes 40-inch instrument; its aperture (front lens) is 40 inches in diameter. The tube is a bit longer.

There are several drawbacks to this design, which are given below.

1. The lens is supported only around the edges; as a result, the glass can sag under the force of gravity, distorting the precise curve of the surfaces and blurring the image.

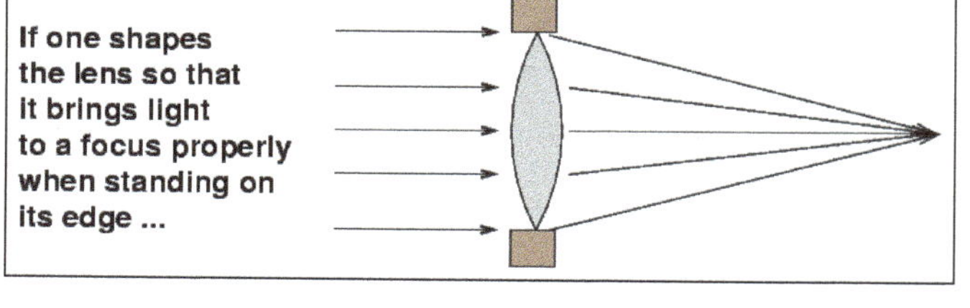

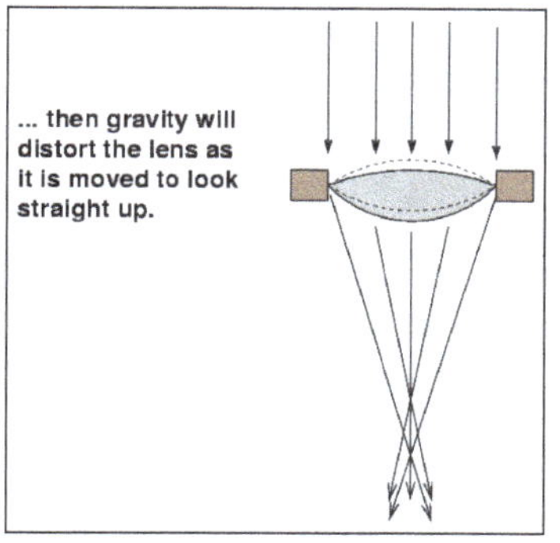

The largest refracting telescope ever built had a lens of diameter 1.25 meters. It was fixed in place to avoid sagging, pointed horizontally at a flat mirror; the flat mirror was tilted to change the part of the sky viewed by the big telescope. It still didn't work.

2. Light of different wavelengths comes to a focus in different places.

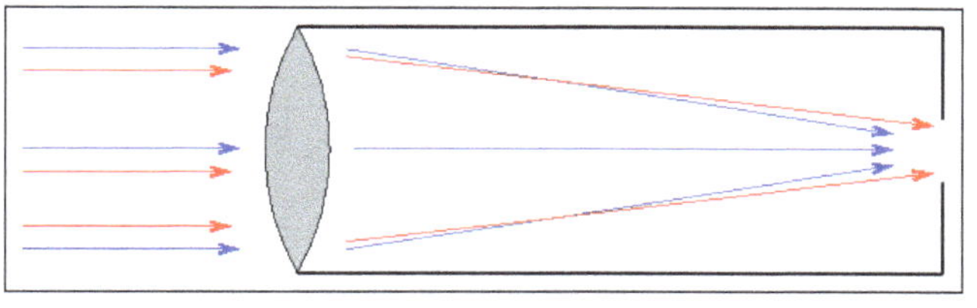

This is called chromatic aberration; it causes bright sources of light to have blue edges on one side and red edges on the other.

Isaac Newton realized that mirrors would solve the second problem: they can bring light of ALL wavelengths to a common focus. He designed a telescope which used mirrors to reflect light; hence, this type is called a reflector.

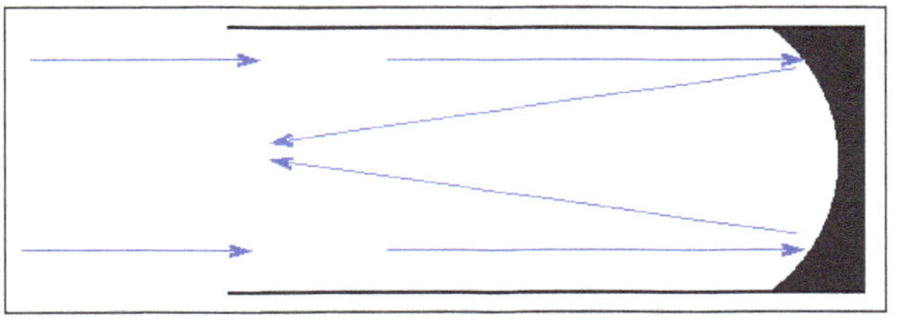

There's a problem with reflectors: they bring light to a focus directly in front of the primary mirror, which isn't such a good idea.

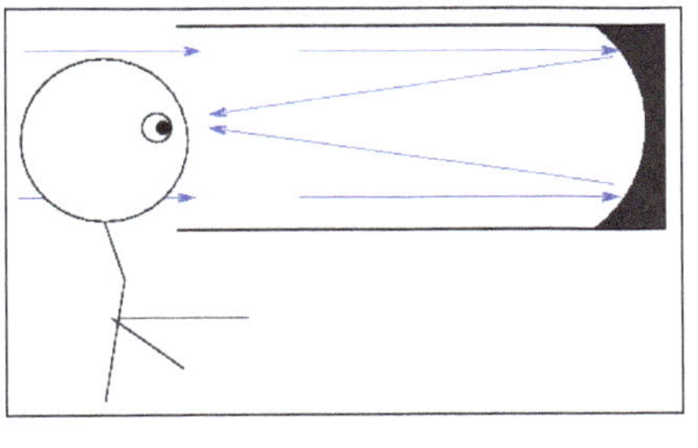

Newton avoided this problem by inserting a *secondary mirror* near the top end of the tube; the flat secondary simply directs the focusing light rays out the side of the tube, where one can place an eyepiece or instrument without blocking the aperture. The secondary mirror does block a little of the incoming light, but the tradeoff is worth it.

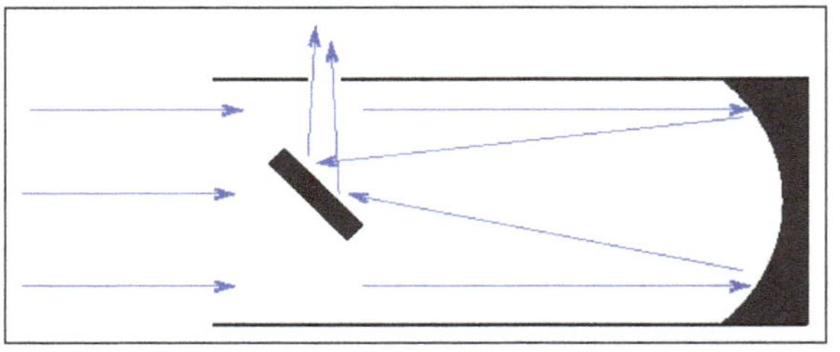

Another common design, called *Cassegrain* or (with a front collecting plate) *Schmidt-Cassegrain*, uses the secondary mirror to send light back down through a hole in the center of the primary mirror.

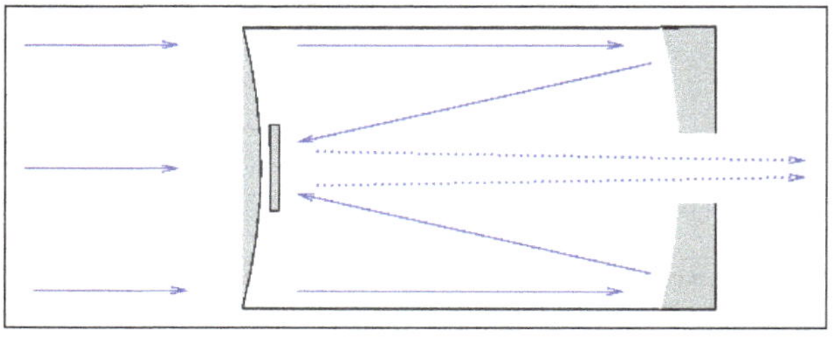

The mirror in a reflector can be supported not only around the edges, but also all over the back surface. That means that very large mirrors can be placed into telescopes. The Subaru Telescope has a primary mirror over 8 meters in diameter.

Larger Aperture has Better Resolution

Astronomers use the word "resolution" to mean "the sharpness of an image"; in other words, the degree to which small details can be discerned. Light behaves like a wave, and one aspect of waves is that they spread out after going past an obstruction -- such as the aperture of a telescope. This diffraction depends on both the wavelength of the light waves, and on the size of the aperture. In general, the smallest angular size a telescope can resolve is:

$$\text{in general,} \qquad \theta = 1.22 \,\text{radians} \frac{\lambda}{\text{D}}$$

where θ is the angular size in radians, λ is the wavelength, and D is the telescope diameter. If we choose a wavelength in the middle of the visible range, λ = 500 nm and convert from radians to arcseconds, and express the diameter D in meters, then we find:

$$\text{in visible} \qquad \theta \sim 0.13 \,\text{arc sec} \frac{1\,\text{meter}}{\text{D}}$$

What is the diffraction limit for an 8-inch telescope? In theory, then, even a modest little telescope should be able to reveal very fine details within lunar craters and Jovian cloud belts.

Unfortunately, it doesn't work that way in practice. The Earth's atmosphere is constantly moving, and different layers bend the light from a star in different directions, blurring our view from the ground.

To a rough approximation, the Earth's atmosphere blurs everything to a minimum resolution of about 1 arcsecond (slightly less at good sites, slightly more at poor ones like Rochester). The picture below shows an example of the atmosphere's effect on telescopic images: it shows a view of the star Betelgeuse from the 4.2-meter William Herschel Telescope.

In order to see really fine detail in the optical, you have to move your telescope above the atmosphere.

The real reason astronomers want big telescopes is to detect fainter and fainter objects. The faintest object visible in a telescope depends on the amount of light the telescope can gather, which in turn depends on its collecting area:

Area = pi * (radius)

Note that the collecting area increases as the square of the radius (or diameter).

The pupil in your eye has a diameter of about 5 mm. The little telescope at the front of the class has a diameter of about 4.25 inches = 108 mm.

The Pleiades is a pretty cluster of stars you can just barely see with your naked eye (picture on left); with the small 4.25-inch telescope mentioned above, you see many more stars.

At some point, however, even though you can support your big mirror across its entire back surface, the mirror becomes so big and so heavy that it starts to sag. Even worse, as you move your telescope to point to different regions of the sky, your mirror will sag in different ways. The mirror will no longer bring all the light rays to a single focus. Our current technology has just about reached that point with mirrors of diameter 8 meters.

What can you do if you want to see even fainter objects? Use an array of smaller mirrors. The Keck Telescopes feature segmented primary mirrors which total about 10 meters across.

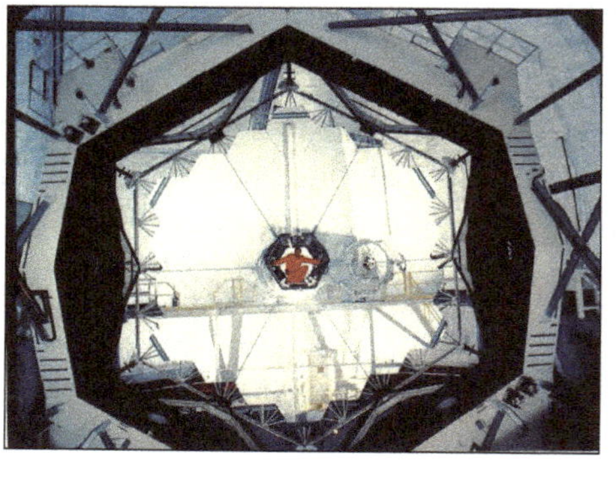

The Hobby-Ebberly Telescope has a slightly larger diameter, though its effective diameter depends slightly on the direction it points.

All the plans to build very large ground-based optical telescopes in the future involve segmented mirrors. Some of these proposed projects are:

- The Giant Magellan Telescope will use seven mirrors, each 8.4 meters in diameter, to collect as much light as a single mirror 24.5 meters across.

- The Thirty Meter Telescope will use 492 individual segments, each about 1.4 meters across, to create an aperture about 30 meters across.

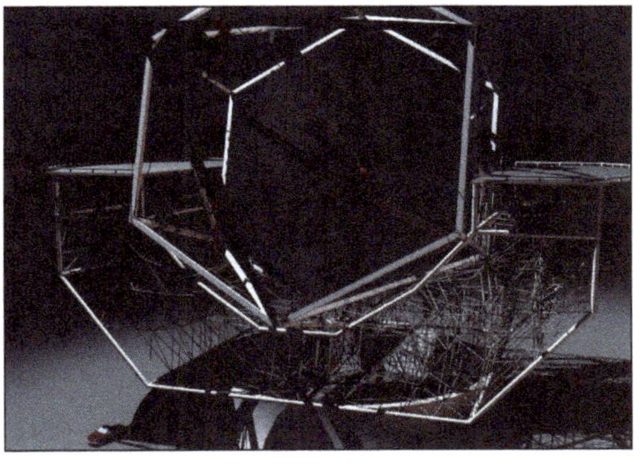

- The Overwhelmingly Large Telescope (OWL) is a European project which plans to build a telescope with an aperture 100 meters across.

Solar Telescopes

Solar telescopes are based on the same construction principles as the night-time instruments, but solar observations require specialized telescopes and instruments, since they must withstand the heat input from the Sun while maintaining their high resolution in the spatial, spectral and temporal dimensions. The solar irradiation heats up the ground, causing a layer of turbulent air which in turn degrades the image quality. Solar telescopes are therefore normally installed in towers, above this turbulent layer.

Most of the existing solar telescopes are synoptic instruments with apertures ranging form a few centimeters to, say, half a meter. Several of these telescopes are organized as networks for helioseismology measurements. Others monitor solar activity and provide images of the solar disk at different wavelength bands, or magnetograms. These telescopes often provide important background information for high-resolution studies, although their importance has somewhat decreased, since the SOHO satellite delivers daily full-disk images without interruptions. New full disk telescopes offer synoptic data at a high cadence, for the investigation of short-lived phenomena.

Telescopes with apertures larger than, say, 0.5 meters have a field of view of only a small fraction of the solar disk at an image scale that allows for diffraction-limited imaging in the focal plane. In the past, most telescopes of the 0.5 to 1 meter class had evacuated light paths in order to suppress inhomogeneities of the air's index of refraction caused inside the telescope by thermal input from the Sun. For solar telescopes of the next generation with apertures of 1.5 to 4 meters, open structures are envisaged, with complex cooling systems for the primary optics that avoid the heating due to absorbed solar radiation. Optical elements are made of material with very low thermal expansion and, if possible, with high thermal conductivity. The latter property simplifies the cooling process and significantly shortens the time needed to reach thermal equilibrium.

Many of the phenomena that can be observed in the solar atmosphere have a lifetime of only a few minutes, and important changes may occur within a few seconds.

High-resolution solar telescopes therefore have to provide light levels high enough to achieve a sufficiently high signal-to noise level. This is most important for the measurement of the (weak) magnetic field in the solar photosphere. Important small-scale objects have sizes of 100 km or less, and it requires telescopes with an aperture of at least one meter to resolve them. Next-generation telescopes with apertures of about four meters will be able to achieve high light level, short integration time, and good spatial resolution. It should be noted that for diffraction-limited observations, the light level per resolution element is the same for any telescope. For an increased light level one therefore has to sacrifice spatial or temporal resolution.

Best Sites

High-quality solar observations require sites with low levels of local and high-altitude turbulence. The atmosphere should also contain little water vapor and dust particles, in order to minimize the amount of scattered light. Sites on high mountains located on rather small islands have proven to be the best solar sites. Low levels of local turbulence can also be obtained at lake sites, where the nearby water keeps the ambient air temperature fluctuations low and inhibits the build-up of local turbulence. The comprehensive ATST solar site site survey, arguably the most testing so far, identifies three excellent site for solar observations: Mees Solar Observatory on Hawaii, Observatorio del Roque de los Muchachos on La Palma and Big Bear Solar Observatory in California. There is evidence that Antarctic sites such as Concordia Station at Dome C might have also have excellent day-time seeing. In the future, the quality of solar telescope sites may be more precisely characterized by the number and altitude of turbulence layers in the atmosphere above the telescope. Multi-conjugated adaptive optics systems will be able to correct the image degradation caused by such well-defined layers.

Existing Telescopes

A large number of solar telescopes with apertures between 150 cm and about 10 cm is presently operational around the globe. Many of the small-aperture telescopes are either used for routine observations of the full disk, or are organized in networks for helioseismic measurements. Three of the large-aperture telescopes are presently equipped with adaptive optics, and are therefore suited for observations with highest possible spatial resolution, for imaging and spectroscopy. The Dunn Solar Telescope, the German Vacuum Tower Telescope, and the Swedish 1-meter Solar Telescope have a number of common features, but also important differences. All three telescopes (i) are tower constructions with the telescope entrance high above ground, above the local layer of turbulence, (ii) have a long focal length of the primary mirror or lens, to avoid a hot focal plane, (iii) use evacuated tubes for the light path, and (iv) are domeless or with retractable dome. The DST and the SST have a altitude-azimuth feed system ("Turret") that allow to have the full light path in vacuum, whereas the VTT uses a Coelostat system. The SST is a refractor, with a 1-m lens that acts also as entrance window.

Next Generation Solar Telescopes

At present, three solar telescopes of the 1.5 to 2 meter class are in preparation or under construction, and two of them should become operational within the next one or two years. These telescopes mark an important design change, since they do no longer rely on evacuated or helium-filled telescope tubes to avoid turbulent air in the light path. They represent an intermediate step between the presently available telescopes and the next-generation 4 meter telescopes. The next generation of solar telescopes with apertures in the range of 4 meters have been enabled by two technical breakthroughs: adaptive optics for solar telescopes, and the feasibility of air-cooled, open telescopes. The Dutch Open Telescope (DOT) on La Palma has been a pathfinder for the new generation of open telescopes.

The German *GREGOR telescope* has an aperture of 1.50 m and is located at the Observatorio del Teide on Tenerife. It is an open telescope in a three-mirror Gregory configuration with a focal length of 50 m. The primary mirror is made of CESIC, a silicon-carbide material with high thermal conduction, and it is air-cooled from the backside. At Big Bear Solar Observatory, the *New Solar Telescope* is under construction. It is an open off-axis Gregory system with an aperture of 160 cm and an effective focal length of 88 m. Both telescopes will be equipped with high-order adaptive optics and become operational before 2010.

In the US, the *Advanced Technology Solar Telescope* (ATST) project of the National Solar Observatory is ready to enter in the constructions phase. The construction phase is expected to start in 2009, and First Light may occur in 2014. The ATST is a 4-meter, off-axis telescope, and it will be constructed on the Haleakala mountain (3000 m) on Hawaii. The telescope design is optimized for high sensitivity, polarimetric accuracy and low scattered light. Due to its open design, the telescope covers a wavelength range from 0.3 µm to 35 µm. The COronal Solar Magnetism Observatory (COSMO), a coronagraph with an aperture of 1.5 meters has been proposed by the High Altitude Observatory in Boulder, and the Universities of Hawaii and Michigan. Phase-A studies for this project are currently underway. In 2007, the European Association for Solar Telescope (EAST) has initiated the *European Solar Telescope* (EST) project. EST is a telescope of the 4-meter class, to be built in the Canary Islands toward the end of the second decade. During a design study, carried out between 2008 and 2010, the opto-mechanical design of EST will be worked out, and a local site characterization will be made. EST will measure the solar magnetic field with highest spatial and spectral resolution in the visible and near infrared wavelength region.

Instrumentation

Adaptive Optics

The recent development of real-time adaptive optics systems to measure and stabilize image motion and to compensate low- and high-order image aberrations led to a major breakthrough in the spatial resolution of solar observations. To date, several telescopes of the 70 – 100 cm class are equipped with adaptive optics systems. These systems can correct atmospheric disturbances with a bandwidth of up to 100 Hz and are capable of correcting the dominant aberration modes caused by the turbulent Earth atmosphere and the instrument itself. The number of aberration modes that can be corrected grows with the number of sub-apertures of the wave-front sensor. The typical size of sub-apertures of a high-order adaptive optics is about 8 cm. This is small enough to account for the anisoplanatism of the day-time atmosphere, but also large enough to resolve the solar photospheric granulation. The area that can be corrected with adaptive optics is very small, only a few arcseconds in diameter. In order to overcome this limitation, multi-conjugated adaptive optics systems are presently under development. Theses systems use several deformable mirrors to correct the wavefront deformations that occur in different heights above the telescope.

The importance and the complexity of adaptive optics for solar observations grows rapidly with increasing telescope aperture. The achievable spatial resolution of the planned telescopes in the U.S. and Europe with apertures in the order of 4 meters will depend critically on the quality of their adaptive optics systems. A high-order system will require wave-front sensors with about 2000 sub-apertures – quite a technical challenge. Fortunately, computing power has grown more rapidly than the size of telescopes, therefore such high-order systems are nowadays within reach.

Filtergraphs

Observations of the smallest details on the Sun, near the diffraction limit of the telescope are made with broad-band imagers. They may consist of a filter to select the wavelength band, and a suitable digital detector, e.g. a CCD camera. Thanks to the high light level, exposure times of a few milliseconds are sufficient. This allows collecting bursts of images in rapid sequence and then selecting the very best ones afterwards, or use the full burst for post-facto image restoration using techniques based on multi-frame blind deconvolution or speckle interferometry. These techniques allow the study of the morphology of the solar photosphere and the evolution of large- and small-scale objects on time scales of a seconds, minutes or longer. Without image restoration, the high-quality field of view of filtergrams is limited by the corrected field (isoplanatic area) of the adaptive optics. However, in practice the full field of view of the available CCDs can be restored to homogeneous quality.

Spectroscopic Instruments

Spectroscopic instruments are needed to obtain physical parameters, such as temperature, magnetic field, or flow speed. These measurements are multi-dimensional: two spatial dimensions, wavelength, and time. At present, detectors can only record two dimensions at a time. There are two different solutions to obtain spectroscopic data: filter instruments that record two-dimensional images at a fixed wavelength, and longslit spectrographs that record one spatial dimension and a certain wavelength range. Both types of instruments have obvious advantages and also disadvantages, and it depends on the scientific topic, which one is preferred. Some solar observatories therefore provide both instruments.

Filter spectrometers record (nearly) monochromatic images. They use tunable narrow-band filters to select the wavelength. Spatial and wavelength information is recorded by taking a sequence of monochromatic images with varying wavelength. Tunable filters can be Lyot filters, or Fabry-Pérot Interferometers or Michelson Interferometers. With a combination of two or three tunable high-quality Fabry-Pérots, a spectral resolution of 2.5 pm can be obtained. The global tuning range is around 300 nm. Due to the small free spectral range, the spectral coverage for an individual measurement is limited to 0.3 nm. Filter spectrometers are often equipped with an additional broad-band channel that takes images in a fixed wavelength band, and simultaneous with the narrowband images. The broad-band sequences are then used for post-facto reconstruction of the data. A typical data set with a field of view of about one arc minute squared and 15 wavelength positions across a spectral line can be taken in a few seconds. The spatial resolution of such a measurement depends on the size of the telescope and the image scale on the detector. Different parts of the spectral line are measured at different times. During times of variable seeing, the shape of the line profile may become distorted.

Long-slit grating spectrographs provide instantaneous information about a certain wavelength range and one spatial dimension (along the slit). The spectral resolution depends mainly on the (illuminated) area of the diffraction grating and the focal length of the instrument. Compact spectrographs have a resolution of 2.5 pm (Resolving power of 250.000), similar to the best filter spectrometers. Spectrographs with large gratings and long focal lengths, like the Echelle spectrograph of the German VTT, have a theoretical resolving power of 1.000.000. Slit spectrographs record one or several spectral lines at a time. This is important for the investigation of the shape of line profiles, because they are not distorted by possible changes in the Earth atmosphere. Two-dimensional spatial information is collected by moving the solar image across the slit. The time needed to cover a certain area depends on the desired spatial resolution, i.e., the slit width and the step size. Fast cadences with high spectral resolution and coverage are possible for small scan areas. Grating spectrographs cover a large range of wavelengths, typically from 380 to 2200 Nanometers.

Spectro-polarimeters are used for the measurement of magnetic field in the solar atmosphere. They exist as combination of filter spectrometers or long-slit spectrographs with suitable polarization modulation components. Since the fraction of polarized light from the Sun is often very small, the needed accuracy of polarimetric measurements is very high. The magnetic signal is obtained by measuring the Stokes parameters that provide information about the total intensity, the circular and two orthogonal states of linear polarization. The polarization modulation is performed either with rotating retarding wave plates, or with modern tunable liquid crystal retarders. A single magnetic field measurement requires at least four different images at different settings of the polarization modulator. In order to minimize the influence of variable seeing conditions, these images have to be taken in rapid sequence. In addition, precise calibrations of the polarization properties of the telescope and the spectro-polarimeter itself are necessary, to guarantee high polarimetric accuracy of the data. The best instruments have an accuracy of 1 part in 10,000.

Radio Telescope

Radio telescope is an astronomical instrument consisting Radio telescope, astronomical instrument consisting of a radio receiver and an antenna system that is used to detect radio-frequency radiation between wavelengths of about 10 metres (30 megahertz [MHz]) and 1 mm (300 gigahertz [GHz]) emitted by extraterrestrial sources, such as stars, galaxies, and quasars.

Extraterrestrial radio emission was first reported in 1933 by Karl Jansky, an engineer at the Bell Telephone Laboratories, while he was searching for the cause of shortwave interference. Jansky had mounted a directional radio antenna on a turntable so that he could point it at different parts of the sky to determine the direction of the interfering signals. He not only detected interference from distant thunderstorms but also located a source of radio "noise" from the centre of the Milky Way Galaxy. This first detection of cosmic radio waves received much attention from the public but only passing notice from the astronomical community.

Grote Reber, a radio engineer and amateur radio operator, built a 9.5-metre parabolic reflector in his backyard in Wheaton, Illinois, U.S., to continue Jansky's investigation of cosmic radio noise. In 1944 he published the first radio map of the sky. After World War II ended, the technology that had been developed for military radar was applied to astronomical research. Radio telescopes of increasing size and sophistication were built first in Australia and Great Britain and later in the United States and other countries.

Principles of Operation

Radio telescopes vary widely, but they all have two basic components: (1) a large radio antenna and (2) a sensitive radiometer, or radio receiver. The sensitivity of a radio telescope—i.e., the ability to measure weak sources of radio emission—depends both on the area and efficiency of the antenna and on the sensitivity of the radio receiver used to amplify and to detect the signals. For broadband continuum emission over a range of wavelengths, the sensitivity also depends on the bandwidth of the receiver. Because cosmic radio sources are extremely weak, radio telescopes are usually very large—up to hundreds of metres across—and use the most sensitive radio receivers available. Moreover, weak cosmic signals can be easily masked by terrestrial radio interference, and great effort is taken to protect radio telescopes from man-made emissions.

The most familiar type of radio telescope is the radio reflector consisting of a parabolic antenna, which operates in the same manner as a television satellite dish to focus the incoming radiation onto a small antenna called the feed, a term that originated with antennas used for radar transmissions. This type of telescope is also known as the dish, or filled-aperture, telescope. In a radio telescope the feed is typically a waveguide horn and transfers the incoming signal to the sensitive radio receiver. Solid-state amplifiers that are cooled to very low temperatures to reduce significantly their internal noise are used to obtain the best possible sensitivity.

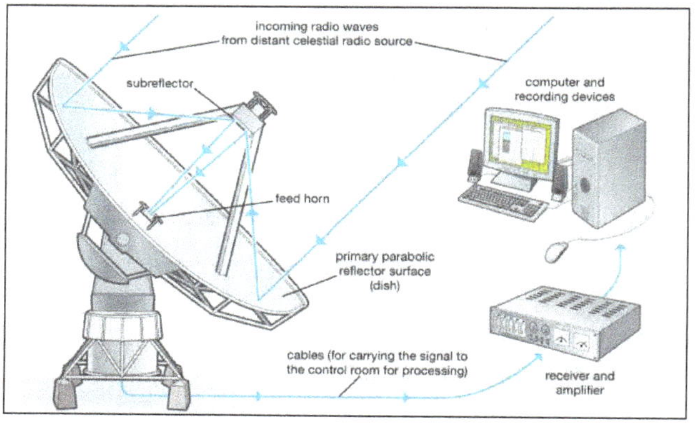

In some radio telescopes the parabolic surface is equatorially mounted, with one axis parallel to the rotation axis of Earth. Equatorial mounts are attractive because they allow the telescope to follow a position in the sky as the Earth rotates by moving the antenna about a single axis parallel to the Earth's axis of rotation. But equatorially mounted radio telescopes are difficult and expensive to build. In most modern radio telescopes, a digital computer is used to drive the telescope about the azimuth and elevation axes to follow the motion of a radio source across the sky.

In the simplest form of radio telescope, the receiver is placed directly at the focal point of the parabolic reflector, and the detected signal is carried by cable along the feed support structure to a point near the ground where it can be recorded and analyzed. However, it is difficult in this type of system to access the instrumentation for maintenance and repair, and weight restrictions limit the size and number of individual receivers that can be installed on the telescope. More often, a secondary reflector is placed in front of (Cassegrain focus) or behind (Gregorian focus) the focal point of the paraboloid to focus the radiation to a point near the vertex, or centre, of the main reflector. Multiple feeds and receivers may be located at the vertex where there is more room, where weight restrictions are less stringent, and where access for maintenance and repair is more straightforward. Secondary focus systems also have the advantage that both the primary and secondary reflecting surfaces may be carefully shaped so as to improve the gain over that of a simple parabolic antenna.

Earlier radio telescopes used a symmetric tripod or quadrapod structure to hold the feed or secondary reflector, but such an arrangement blocks some of the incoming radiation, and the reflection of signals from the support legs back into the receiver distorts the response. In newer designs, the feed or secondary reflector is placed off the central axis and does not block the incoming signal. Off-axis radio telescopes are thus more sensitive and less affected by interference reflected from the support structure into the feed.

The performance of a radio telescope is limited by various factors. The accuracy of a reflecting surface may depart from the ideal shape because of manufacturing irregularities. Wind load can exert force on the telescope. Thermal deformations cause differential expansion and contraction. As the antenna is pointed to different parts of the sky, deflections occur due to changes in gravitational forces. Departures from a perfect parabolic surface become important when they are a few percent or more of the wavelength of operation. Since small structures can be built with greater precision than larger ones, radio telescopes designed for operation at millimetre wavelengths are typically only a few tens of metres across, whereas those designed for operation at centimetre wavelengths range up to 300 metres (1,000 feet) in diameter. For operation at relatively long metre wavelengths where the reflecting surface need not have an accuracy better than a few centimetres, it becomes practical to build very large fixed structures in which the reflecting surface can be made of simple "chicken wire" fencing or even parallel rows of wires.

Traditionally the effect of gravity has been minimized by designing the movable structure to be as stiff as possible in order to reduce the deflections resulting from gravity. A more effective technique, based on the principle of homology, allows the structure to deform under the force of gravity, and the cross section and weight of each member of the movable structure are chosen to cause the gravitational forces to deform the reflecting structure into a new paraboloid with a slightly different focal point. It is then necessary only to move the feed or secondary reflector to maintain optimum performance. Homologous designs have become possible only since the development of computer-aided structural simulations known as the finite element method.

Some radio telescopes, particularly those designed for operation at very short wavelengths, are placed in protective enclosures called radomes that can nearly eliminate the effect of both wind loading and temperature differences throughout the structure. Special materials that exhibit very low absorption and reflection of radio waves have been developed for such structures, but the cost of enclosing a large antenna in a suitable temperature-controlled radome may be almost as much as the cost of the movable antenna itself.

The 20-metre (60-foot) diameter,
radome-enclosed, millimetre-wave telescope.

The cost of constructing an antenna with a very large aperture can be greatly reduced by fixing the structure to the ground and moving either the feed or the secondary reflector to steer the beam in the sky. However, for parabolic reflecting surfaces, the beam can be steered in this way over only a limited range of angle without introducing aberration and a loss of signal strength.

Radio telescopes are used to measure broad-bandwidth continuum radiation as well as narrow-bandwidth spectroscopic features due to atomic and molecular lines found in the radio spectrum of astronomical objects. In early radio telescopes, spectroscopic observations were made by tuning a receiver across a sufficiently large frequency range to cover the various frequencies of interest. Because the spectrometer had a narrow frequency range, this procedure was extremely time-consuming, and it greatly restricted observations. Modern radio telescopes observe simultaneously at a large number of frequencies by dividing the signals up into as many as several thousand separate frequency channels that can range over a much larger total bandwidth of tens to hundreds of megahertz.

The most straightforward type of radio spectrometer employs a large number of filters, each tuned to a separate frequency and followed by a separate detector that combines the signal from the various filters to produce a multichannel, or multifrequency, receiver. Alternatively, a single broad-bandwidth signal may be converted into digital form and analyzed by the mathematical process of autocorrelation and Fourier transforms. In order to detect faint signals, the receiver output is often averaged over periods of up to several hours to reduce the effect of noise generated by thermal radiation in the receiver.

Radio Interferometry and Aperture Synthesis

The angular resolution, or ability of a radio telescope to distinguish fine detail in the sky, depends on the wavelength of observations divided by the size of the instrument. Yet even the largest antennas, when used at their shortest operating wavelength, have an angular resolution of only a few arc seconds, which is about 10 times poorer than the resolution of ground-based optical telescopes. Because radio telescopes operate at much longer wavelengths than do optical telescopes, radio telescopes need to be much larger than optical telescopes to achieve the same angular resolution.

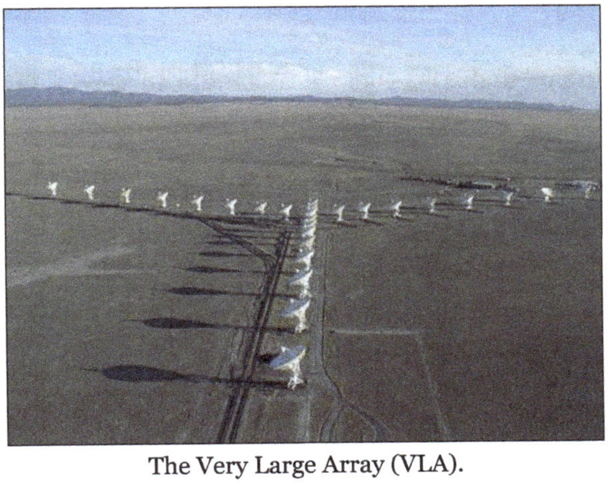

The Very Large Array (VLA).

At radio wavelengths, the distortions introduced by the atmosphere are less important than at optical wavelengths, and so the theoretical angular resolution of a radio telescope can in practice be achieved even for the largest dimensions. Also, because radio signals are easy to distribute over large distances without distortion, it is possible to build radio telescopes of essentially unlimited dimensions. In fact, the history of radio astronomy has been one of solving engineering problems to construct radio telescopes of continually increasing angular resolution.

The high angular resolution of radio telescopes is achieved by using the principles of interferometry to synthesize a very large effective aperture from a number of antennas. In a simple two-antenna radio interferometer, the signals from an unresolved, or "point," source alternately arrive in phase (constructive interference) and out of phase (destructive interference) as Earth rotates and causes a change in the difference in path from the radio source to the two elements of the interferometer. This produces interference fringes in a manner similar to that in an optical interferometer. If the radio source has finite angular size, then the difference in path length to the elements of the interferometer varies across the source. The measured interference fringes from each interferometer pair thus depend on the detailed nature of the radio "brightness" distribution in the sky.

Each interferometer pair measures one "Fourier component" of the brightness distribution of the radio source. Work by Sir Martin Ryle and his colleagues in the1950s and '60s showed that movable antenna elements combined with the rotation of Earth can sample a sufficient number of Fourier components with which to synthesize the effect of a large aperture and thereby reconstruct high-resolution images of the radio sky. The laborious computational task of doing Fourier transforms to obtain images from the interferometer data is accomplished with high-speed computers and the fast Fourier transform (FFT), a mathematical technique that is specially suited for computing discrete Fourier transforms. In recognition of his contributions to the development of the Fourier synthesis technique, more commonly known as aperture synthesis, or earth-rotation synthesis, Ryle was awarded the 1974 Nobel Prize for Physics.

During the 1960s the Swedish physicist Jan Hogbom developed a technique called CLEAN, which is used to remove the spurious responses from a celestial radio image caused by the use of discrete, rather than continuous, spacings in deriving the radio image. Further developments, based on a technique introduced in the early 1950s by the British scientists Roger Jennison and Francis Graham Smith, led to the concept of self-calibration, which uses the observed source as its own calibrator in order to remove errors in a radio image due to uncertainties in the response of individual antennas as well as small errors introduced by the propagation of radio signals through the terrestrial atmosphere. In this way radio telescopes are able to achieve extraordinary angular resolution and image quality that are not possible in any other wavelength band.

Very Long Baseline Interferometry

In conventional interferometers and arrays, coaxial cable, waveguide, or even fibre-optic links are used to distribute a common local-oscillator reference signal to each antenna and also to return the received signal from an individual antenna to a central laboratory where it is correlated with the signals from other antennas. In cases in which antennas are spaced more than a few tens of kilometres apart, however, it becomes prohibitively expensive to employ real physical links to distribute the signals. Very high frequency (VHF) or ultrahigh frequency (UHF) radio links have been used, but the need for a large number of repeater stations makes this impractical for spacings greater than a few hundred kilometres.

Interferometer systems of essentially unlimited element separation can be formed by using the technique of very long baseline interferometry (VLBI). In early VLBI systems the signals received at each element were recorded by broad-bandwidth videotape recorders located at each antenna. More recently, with the advent of inexpensive, reliable computer disk drives, the data are recorded on disks. The disks are then transported to a common location where they are replayed and the signals combined to form interference fringes. The successful operation of a VLBI system requires that the tape recordings be synchronized within a few millionths of a second and that the local oscillator reference signal be stable to better than one part in a trillion. Recorded data from just a few hours of observation typically contain about one trillion bits of information, which is roughly equivalent to storing the entire contents of a modest-sized library. Hydrogen maser frequency standards are used to give a timing accuracy of only a few billionths of a second and a frequency stability of one part in a million billion.

Radar Techniques

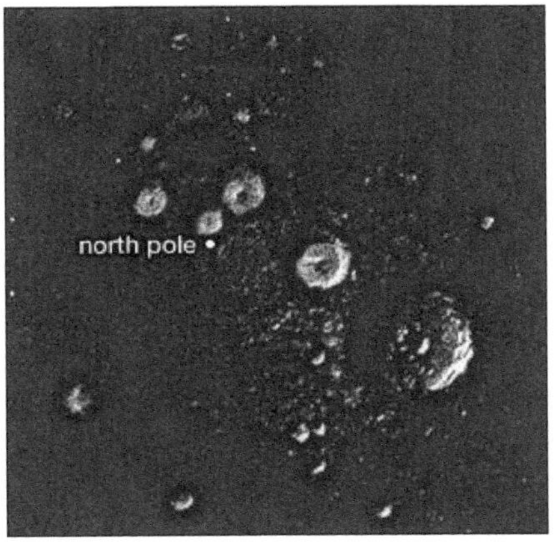

Techniques analogous to those used in military and civilian radar applications are sometimes employed with radio telescopes to study the surface of planets and asteroids in the solar system. By measuring the spectrum and the time of flight of signals reflected from planetary surfaces, it is possible to examine topographical features with a linear resolution as good as 1 km, deduce rates of rotation, and determine with great accuracy the distance to the planets. Radio signals reflected from the planets are weak, and high-power radar transmitters are needed in order to obtain measurable signal detections. The time it takes for a radar signal to travel to Venus and back, even at the closest approach of the planet to Earth, is about five minutes. For Saturn, it is more than two hours.

Mercury's north polar region, in a radar image obtained with the Arecibo radio telescope. All the bright (radar-reflective) features are believed to be deposits of frozen volatile substances, likely water ice, at least several metres thick in the permanently shaded floors of craters.

Major Applications of Radio Telescopes

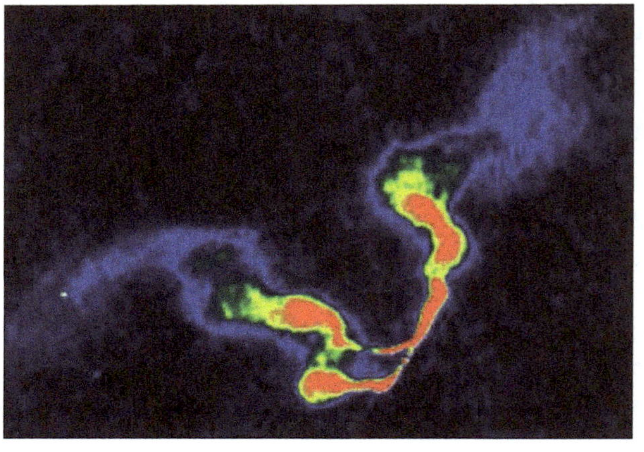

Radio telescopes permit astronomers to study many kinds of extraterrestrial radio sources. These astronomical objects emit radio waves by one of several processes, including (1) thermal radiation from solid bodies such as the planets, (2) thermal, or bremsstrahlung, radiation from hot gas in the interstellar medium, (3) synchrotron radiation from electrons moving at velocities near the speed of light in weak magnetic fields, (4) spectral line radiation from atomic or molecular transitions that occur in the interstellar medium or in the gaseous envelopes around stars, and (5) pulsed radiation resulting from the rapid rotation of neutron stars surrounded by an intense magnetic field and energetic electrons.

Image of the radio source 3C 75 in the cluster of galaxies Abell 400 taken with the Very Large Array (VLA) at Socorro, New Mexico, at a wavelength of 20 cm (8 inches). Red shows regions of intense radio emission, while blue shows regions of fainter emission. The image consists of two twin jet radio sources. The jets bend and appear to be interacting.

Radio telescopes are used to measure the surface temperatures of all the planets, as well as some of the moons of Jupiter and Saturn. Radar measurements have revealed the rotation of Mercury, which was previously thought to keep the same side toward the Sun. Astronomers have also used radar observations to image features on the surface of Venus, which is completely obscured from visual scrutiny by the heavy cloud cover that permanently enshrouds the planet. Accurate measurements of the travel time of radar signals reflected from Venus when it is on the other side of the Sun from Earth have indicated that radio waves passing close to the Sun slow down owing to gravity and thereby provide a new independent test of Albert Einstein's theory of general relativity.

Broadband continuum emission throughout the radio-frequency spectrum is observed from a variety of stars (especially binary, X-ray, and other active stars), from supernova remnants, and from magnetic fields and relativistic electrons in the interstellar medium. The discovery of pulsars (short for *pulsating radio stars*) in 1967 revealed the existence of rapidly rotating neutron stars throughout the Milky Way Galaxy and led to the first observation of the effect of gravitational radiation.

Using radio telescopes equipped with sensitive spectrometers, radio astronomers have discovered about 150 separate molecules, including familiar chemical compounds such as water, formaldehyde, ammonia, methanol, ethyl alcohol, and carbon dioxide. The important spectral line of atomic hydrogen at 1,421.405 MHz (21-cm wavelength) is used to determine the motions of hydrogen clouds in the Milky Way Galaxy and other galaxies. This is done by measuring the change in the wavelength of the observed lines arising from the Doppler effect. It has been established from such measurements that the rotational velocities of the hydrogen clouds vary with distance from the galactic centre. The mass of a spiral galaxy can in turn be estimated using this velocity data. In this way radio telescopes show evidence for the presence of so-called dark matter by showing that the amount of starlight is insufficient to account for the large mass inferred from the rapid rotation curves.

Radio telescopes have discovered powerful radio galaxies and quasars far beyond the Milky Way Galaxy system. These cosmic objects have intense clouds of radio emission that extend hundreds of thousands of light-years away from a central energy source located in an active galactic nucleus (AGN), or quasar. Observations with high-resolution radio arrays show highly relativistic jets extending from an AGN to the radio lobes.

Measurements made in 1965 by Arno Penzias and Robert Wilson using an experimental communications antenna at 3-cm wavelength located at Bell Laboratories in Holmdel, New Jersey, detected the existence of a cosmic microwave background radiation with a temperature of 3 kelvins (K). This radiation, which comes from all parts of the sky, is thought to be the remaining radiation from the hot big bang, the primeval explosion from which the universe presumably originated 13.7 billion years ago. Satellite and ground-based radio telescopes have been used to measure the very small deviations from isotropy of the cosmic microwave background. This work has led to refined determination of the size, geometry, and age of the universe.

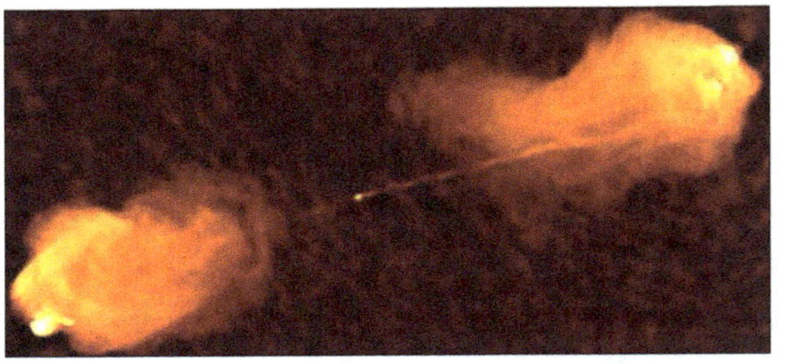

The powerful radio galaxy Cygnus A. The radio waves are coming from electrons propelled at nearly the speed of light through a long, thin jet at the core of the galaxy and deposited in the giant lobes.

Hubble Space Telescope

The Hubble Space Telescope is a large telescope in space. It was launched into orbit by space shuttle Discovery on April 24, 1990. Hubble orbits about 547 kilometers (340 miles) above Earth. It is the length of a large school bus and weighs as much as two adult elephants. Hubble travels about 5 miles per second: That is like traveling from the eastern coast of the United States to the western coast in 10 minutes. Hubble is solar-powered.

Hubble takes sharp pictures of objects in the sky such as planets, stars and galaxies. Hubble has made more than one million observations. These include detailed pictures of the birth and death of stars, galaxies billions of light years away, and comet pieces crashing into Jupiter's atmosphere.

Scientists have learned a lot about the universe from these pictures. Many of them are beautiful to look at.

The Hubble Space Telescope makes one orbit around Earth every 95 minutes.

Difference between Hubble and other Telescopes on Earth

Earth's atmosphere alters and blocks the light that comes from space. Hubble orbits above Earth's atmosphere, which gives it a better view of the universe than telescopes have at ground level.

Hubble is named after an American astronomer, Edwin P. Hubble. He made important discoveries in the early 1900s. He showed that the galaxy containing the solar system -- the Milky Way -- was only one of many galaxies. His work helped show that the universe is expanding. This led to the big-bang theory, which says that the universe began with an intense burst of energy and has been expanding ever since.

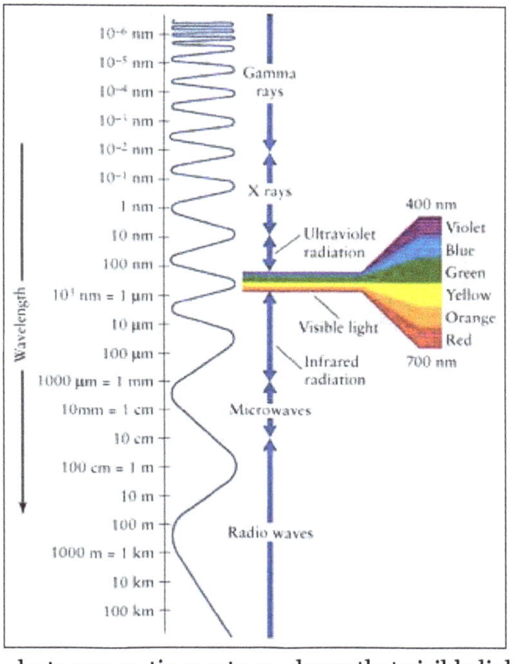

The electromagnetic spectrum shows that visible light is between infrared radiation and ultraviolet radiation.

Instruments on Hubble

As Hubble orbits Earth, the Fine Guidance Sensors lock onto stars. The Fine Guidance Sensors are part of the Pointing Control System and aim Hubble in the right direction. The telescope can lock onto a target that is one mile away without moving more than the width of a human hair.

Once the target is acquired, Hubble's primary mirror collects light. The mirror can collect about 40,000 times more light than the human eye. The light bounces off the primary mirror to the secondary mirror. The secondary mirror focuses the light back through a hole in the primary mirror. From there, the light shines to Hubble's scientific instruments. Each instrument has a different way of interpreting the light.

Hubble has five scientific instruments which include cameras and spectrographs. A spectrograph is an instrument that splits light into its individual wavelengths.

The Wide Field Camera 3 is Hubble's main camera. It studies everything from the formation of distant galaxies to the planets in the solar system. The camera can see three different kinds of light: near-ultraviolet, visible and near-infrared. But Hubble can only see each kind of light one at a time. Human eyes can see visible light. Near-ultraviolet and near-infrared are just beyond what our eyes can see.

The Advanced Camera for Surveys captures images of large areas of space. These images have helped scientists study some of the earliest activity in the universe. The Cosmic Origins Spectrograph reads ultraviolet light. This spectrograph studies how galaxies, stars and planets formed and changed. The Space Telescope Imaging Spectrograph helps scientists determine the temperature, chemical composition, density and motion of objects in space. It also has been used to detect black holes.

The Near Infrared Camera and Multi-Object Spectrometer, or NICMOS, sees objects in deep space by sensing the heat they emit. It captures images and it is also a spectrograph. NICMOS helps scientists study how stars, galaxies and planetary systems form.

The Hubble Space Telescope took this picture of the Tadpole Galaxy and its tail of large, bright blue star clusters.

Important Discoveries

Images taken by Hubble have helped scientists estimate the age and size of the universe. Scientists believe the universe is almost 14 billion years old. Hubble has helped scientists understand how planets and galaxies form. An image called "Hubble Ultra Deep Field" shows the farthest galaxies ever seen.

Hubble has detected black holes, which suck in everything around them, including light. The telescope has played a key role in the discovery of dark energy, a mysterious force that causes the universe to expand faster and faster as time goes on. And it has revealed details of gamma-ray bursts -- powerful explosions of energy that occur when massive stars collapse.

Hubble has also studied the atmospheres of planets revolving around stars similar to Earth's sun.

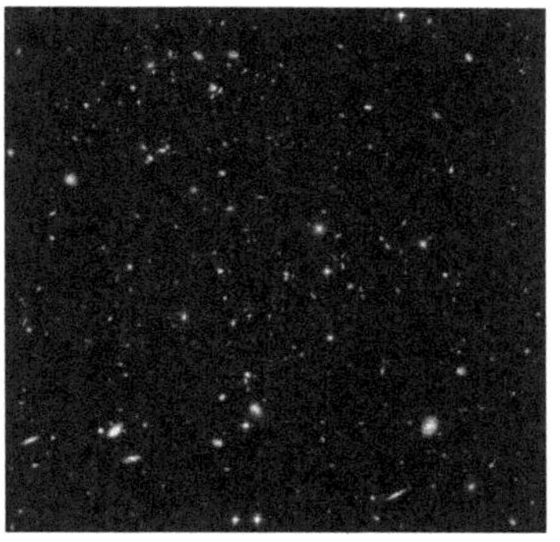

The “Hubble Ultra Deep Field” shows many galaxies far from Earth.

Hubble transmits about 140 gigabytes of science data every week back to Earth. That’s equal to about 45 two-hour, HD-quality movies or about 30,000 mp3 songs. The digital signals are relayed to satellites, then to a ground station, then to NASA’s Goddard Space Flight Center, and finally to the Space Telescope Science Institute. The STScI translates the data into images and information we can understand.

Hubble pictures start out as shades of black and white. The Space Telescope Science Institute adds colors to the pictures for different reasons. Sometimes colors are chosen to show how an object might look to the human eye. Other times colors are used to highlight an important detail. Or they can be used to show details that would otherwise be invisible to the human eye.

Permissions

All chapters in this book are published with permission under the Creative Commons Attribution Share Alike License or equivalent. Every chapter published in this book has been scrutinized by our experts. Their significance has been extensively debated. The topics covered herein carry significant information for a comprehensive understanding. They may even be implemented as practical applications or may be referred to as a beginning point for further studies.

We would like to thank the editorial team for lending their expertise to make the book truly unique. They have played a crucial role in the development of this book. Without their invaluable contributions this book wouldn't have been possible. They have made vital efforts to compile up to date information on the varied aspects of this subject to make this book a valuable addition to the collection of many professionals and students.

This book was conceptualized with the vision of imparting up-to-date and integrated information in this field. To ensure the same, a matchless editorial board was set up. Every individual on the board went through rigorous rounds of assessment to prove their worth. After which they invested a large part of their time researching and compiling the most relevant data for our readers.

The editorial board has been involved in producing this book since its inception. They have spent rigorous hours researching and exploring the diverse topics which have resulted in the successful publishing of this book. They have passed on their knowledge of decades through this book. To expedite this challenging task, the publisher supported the team at every step. A small team of assistant editors was also appointed to further simplify the editing procedure and attain best results for the readers.

Apart from the editorial board, the designing team has also invested a significant amount of their time in understanding the subject and creating the most relevant covers. They scrutinized every image to scout for the most suitable representation of the subject and create an appropriate cover for the book.

The publishing team has been an ardent support to the editorial, designing and production team. Their endless efforts to recruit the best for this project, has resulted in the accomplishment of this book. They are a veteran in the field of academics and their pool of knowledge is as vast as their experience in printing. Their expertise and guidance has proved useful at every step. Their uncompromising quality standards have made this book an exceptional effort. Their encouragement from time to time has been an inspiration for everyone.

The publisher and the editorial board hope that this book will prove to be a valuable piece of knowledge for students, practitioners and scholars across the globe.

Index